This splendidly clear and concise introduction to the craft should be a foundation of every student's apprenticeship—and for those who missed it, a toolkit for a salutary retrofit later. How much more quickly and pleasantly we would discover truth if everyone followed these simple precepts!

Amory B. Lovins
CEO
Rocky Mountain Institute

As information becomes ever cheaper, Dr. Koomey's book becomes ever more valuable. Masterful!

Professor Erik Brynjolfsson
Sloan School of Management, Massachusetts Institute of Technology
Director, Center for E-Business @ MIT
Co-editor, *Understanding the Digital Economy*

Dr. Koomey has produced an absolute home run! A witty, incisive primer on critical analytical thinking. Required reading for business analysts, planners, and strategists—critical insight for players in the Internet economy.

Tod Loofbourrow
President and CEO
Authoria, Inc.

Here at last is the definitive guide for beating information overload and responding to the current anti-science, anti-environment backlash. This remarkable book will empower both professionals and neophytes.

Professor John Harte
Energy and Resources Group, University of California, Berkeley
Author of *Consider a Spherical Cow:*
A Course in Environmental Problem Solving

The greatest challenge facing educators, institutions, and businesses is the inculcation of rigor and fact-based analysis into the psyches of future leaders. *Turning Numbers Into Knowledge* is the tool for organizations to accomplish this essential task.

Ephraim Heller
Founder, Therasense, Inc.

I like this book! Both practical and philosophical, Koomey's work gives readers the skills to extract, communicate, and most importantly to understand knowledge embodied in numbers.

Eric Bergman, Ph.D.
Sun Microsystems
Author of *Information Appliances and Beyond*

Turning Numbers Into Knowledge is an insightful, entertaining, and practical guide to critical thinking in the Information Age.

Professor Stephen J. DeCanio
Department of Economics,
University of California, Santa Barbara

Knowing what numbers really have to say in today's data-intensive world is knowing the difference between fact, fiction, and the yet unknown—and key for personal and professional success. *Turning Numbers into Knowledge* is a must read!

Kathleen Hogan, Ph.D.
Director, Climate Protection Division
U.S. Environmental Protection Agency

Turning Numbers into Knowledge is the long-awaited Rosetta stone between two worlds: that of the undergraduate student aspiring to make a difference in the world, and that of the professional policy analyst working to make sense of it. It will enlighten and empower a whole generation of students who want to change policy and need to know how. **Professor Michael Maniates**
Departments of Political Science and
Environmental Science, Allegheny College

Turning Numbers into Knowledge teaches tricks that most problem solvers only learn after years on the job. Give yourself an edge: Read this book!

Barbara Barkovich, Ph.D.
Principal
Barkovich and Yap, Inc., Consultants

Using numbers to describe the world is an art that takes understanding, skill, and courage. This book supplies the artist with perspective and the practitioner with reliable equipment

Kim Taylor, Ph.D.
Research Program Manager
U.S. Geological Service

Turning Numbers into Knowledge

MASTERING THE ART OF PROBLEM SOLVING

JONATHAN G. KOOMEY, PH.D.

http://www.numbersintoknowledge.com

Analytics Press
PO Box 20313
Oakland, CA 94620-0313
http://www.analyticspress.com

Analytics Press
PO Box 20313
Oakland, CA 94620-0313
SAN 253-5602
Internet: http://www.numbersintoknowledge.com
email: numbers@numbersintoknowledge.com

ISBN 0-9706019-0-5
Library of Congress Card Number 00-193242

Production manager: Susanna Tadlock and David Peattie
Designer: Sandy Drooker
Compositor: BookMatters
Cover art: Tom Chen
Printer and binder: The Maple-Vail Book Manufacturing Group

Text credits:
The graphics characterizing the Cycle of Action in Chapter 3
(Information, intention, and action) are adapted from *The
Psychology of Everyday Things* by Donald A. Norman. © 1988 by
Donald A. Norman. Reprinted by permission of Basic Books, a
member of Perseus Books, L.L.C.

Figure 19.1 is adapted from "Strategies, Structures, and Processes
of Organizational Decisions," which is from *Comparative Studies
in Administration*, T. E. Thompson and A. Tuden, © 1959.
Reprinted by permission of The University of Pittsburgh Press.

Quotations from Peter Schwartz in Chapter 26 (Tell a good story)
taken from The *Art of the Long View* by Peter Schwartz, © 1991
by Peter Schwartz. Used by permission of Doubleday, a division of
Random House, Inc.

Table 36.1 is taken from *A Scientist in the City* by James Trefil, text
© 1994 by James Trefil, illustrations © 1994 by Judith Peatross.
Used by permission of Doubleday, a division of Random House, Inc.

This book is printed on acid-free paper in the United States of America.
10 9 8 7 6 5 4 3

. . . I'd always believed that a life of quality, enjoyment, and wisdom were my human birthright and would be automatically bestowed upon me as time passed. I never suspected that I would have to learn how to live—that there were specific disciplines and ways of seeing the world I had to master before I could awaken to a simple, happy, uncomplicated life. — **DAN MILLMAN**

CONTENTS

FOREWORD

Jon Koomey writes books the old fashioned way: by accumulating experiences, anecdotes and examples over a lifetime of hard work, and refining them into a compact ingot of pure gold. The reader is offered a rare gift indeed—the essential elements of dozens of fine books, the collective wisdom of countless scientists and commentators, and a handful of the most inspired comic strips ever to grace the daily newspaper. In an era where so many books, once read, become disposable, Dr. Koomey has created an enduring reference. His focus on the Internet and his commitment to keeping its content fresh through an ongoing electronic dialogue with readers are laudable indeed. Read this book and apply its many lessons not just to school or work, but to *life*!

Chris Calwell, Founder,
ECOS Consulting, Inc.

PREFACE

Whatever failures I have known, whatever errors I have committed, whatever follies I have witnessed in private and public life have been the consequence of action without thought. —**BERNARD BARUCH**

Quantitative problem solving is the process by which we take *numbers* and transform them into *knowledge*, using our instincts and experience to fill in when we don't have all the answers. Although the technical aspects of this process are taught at many universities, the *art* of problem solving is rarely discussed and even more rarely written down. This book teaches the intricacies of that art and will help you become a first-rate analyst in your chosen field.

After reading this book, you will be well equipped to make independent judgments about analysis used by others. You will know which key questions to ask, so you need never again be at the mercy of those who traffic in "proof by vigorous assertion."[2] You will also be more effective at conducting and presenting your own analyses, no matter what the topic.

Mastering the art of problem solving takes more than proficiency with basic calculations: it requires familiarity with how people use information and requires learning about things as diverse as exploring your ideology, telling good stories, and distinguishing facts from values. To give you a feeling for what to expect, I present an annotated chapter list below.

ANNOTATED CHAPTER LIST

This book contains five major sections, separated into 38 short chapters. Each chapter is compact and self-contained, and each summarizes key lessons I've learned over the years.

INTRODUCTION • THE INFORMATION EXPLOSION: This section briefly describes how analysis can help reduce the information overload that affects us all.

PART I • THINGS TO KNOW: These chapters summarize ideas to keep in mind as you read the rest of the book. More experienced analysts should delve into the ones they find most intriguing and skim the rest.

Chapter 1 • Beginner's Mind: Start fresh and approach any problem like a beginner would, and you'll surely see things that others will miss.

Chapter 2 • Don't Be Intimidated: The difference between success and failure often depends on whether you are intimidated. By consciously refusing to be cowed, you can stack the odds in your favor.

Chapter 3 • Information, Intention, and Action: This chapter describes how humans respond to events, exploring the connections between what we measure, what we assume, and what we choose to do.

Chapter 4 • Peer Review and Scientific Discovery: Progress in science can be subject to human frailty, just as can any other human endeavor. The end result, however, is something you can count on, in large part because of the peer review process.

PART II • BE PREPARED: A key determinant of your effectiveness is the quality of your preparation. Whether you're building a house or chairing a meeting, preparation for the analysis tasks at hand can turn a potential disaster into a triumph.

Chapter 5 • Explore Your Ideology: Ideology provides a simplified model of the world that reflects your values and experiences and prevents paralysis in the face of the many choices you must make every day. Make sure you know your own belief system and those of others.

Chapter 6 • Get Organized: Working and living in chaos is like running a marathon with your feet tied together. Get your life in shape and keep it that way.

Chapter 7 • Establish a Filing System: Few mistakes are more maddening than knowing you have seen a relevant article and not being able to find it. By creating a good filing system, you can prevent this annoyance from ever happening again.

Chapter 8 • Build a Toolbox: My analytical toolbox is the set of tricks and techniques that I can pull out to solve particular problems. This chapter describes some key tools to consider for your own.

Chapter 9 • Put Facts at Your Fingertips: Every analysis requires data. Unless you've memorized the encyclopedia, you'll want to keep some key reference sources within easy reach. This chapter describes the ones I find most useful.

Chapter 10 • Value Your Time: Remember that your time is your life, and that if someone is wasting your time, they are stealing your life. Identify your most productive times of day and protect yourself from interruptions during those periods. Unplug the phone. Go to the library. Take control of those times!

PART III • ASSESS THEIR ANALYSIS: When faced with the assertions of others, it's good to know the right questions to ask. These chapters summarize hard-won knowledge about deciphering other people's analyses.

Chapter 11 • The Power of Critical Thinking: Careful critical thinking is at the root of all good analysis. When the steps described in this chapter become second nature, you will have mastered its essence.

Chapter 12 • Numbers Aren't Everything: Not everything that matters can be quantified, so make sure the unmeasurable doesn't fall through the cracks.

Chapter 13 • All Numbers Are Not Created Equal: Numbers and calculations characterizing the physical world are almost always more certain than those describing human behavior. Many analysts wrongly imply that forecasts based on economic data are just as solid as science. They aren't, so be forewarned.

Chapter 14 • Question Authority: This catch phrase of the 1960s is still applicable today. Authority figures can be wrong or biased, so investigate their assertions in the same way that you'd examine those of someone with whom you're not familiar.

Chapter 15 • How Guesses Become Facts: Always remember that "official" statistics are based on calculations that are often poorly documented, incorrectly cited, or otherwise hazardous to your intellectual health.

Chapter 16 • Don't Believe Everything You Read: Maintain a healthy skepticism, even of well-established sources. In this age of instant information transmission, rumor and error seem to propagate even more quickly than truth.

Chapter 17 • Go Back to the Questions: Any time you rely on survey data to make an important decision, refer back to the actual questionnaire upon which the survey data are based; otherwise, you risk misinterpreting the data.

Chapter 18 • Reading Tables and Graphs: First check for internal consistency, then see if the results contradict other facts you know to be true. Search for cognitive dissonance; any discrepancy between the author's results and what you already know will help you investigate further.

Chapter 19 • Distinguish Facts From Values: Don't be fooled by technical people who portray their advice as totally rational, completely objective, and value-free. If they have made a choice, they have also made a value judgment.

Chapter 20 • The Uncertainty Principle and the Mass Media: Just as the observer of a subatomic particle can disturb that particle by the act of observation, the observer of an institution can disturb that institution by observing and reporting on it. Members of the media (and the analysts who inform them) should take responsibility for the power they wield.

PART IV • CREATE YOUR ANALYSIS: Everyone develops his/her own techniques for creating cogent analyses, and in this section I summarize those I've learned. The importance of organization, clear thinking, careful definitions, systematic exposition, scrupulous documentation, and consistent comparisons cannot be overestimated, and you'll learn about each of these here.

Chapter 21 • Reflect: Free yourself from interruptions and give yourself time to reflect. Without such time, you'll never achieve your full problem-solving potential.

Chapter 22 • Getting Unstuck: Everyone gets stuck sometimes, but this pitfall need not hobble your efforts if you use the tricks in this chapter.

Chapter 23 • Inquire: When faced with a problem outside your expertise, don't surrender! It's an advantage to be unconstrained by the mental shackles most disciplines place on their practitioners. Some of the most important insights in modern thought came from people who could think "outside the box."

Chapter 24 • Be a Detective: Detectives are real-world practitioners of the scientific method. The time-honored techniques of these seasoned problem solvers can help you in your efforts.

Chapter 25 • Create Consistent Comparisons: People often relate best to anecdotes. A consistent comparison is a well-chosen set of two anecdotes that illustrates your point in a compelling way. It is a powerful technique and one well worth learning.

Chapter 26 • Tell a Good Story: Scenario analysis is the art of structured storytelling, and it's an essential tool for any good analyst. Most people don't realize that this art is both highly developed and pertinent to many everyday situations.

Chapter 27 • Dig into the Numbers: Don't be shy about delving into the actual numbers even if you're a highly paid executive. You'll learn things you'd never see if someone else crunches the numbers.

Chapter 28 • Make a Model: Models are "laboratories for the imagination," and this chapter explores the subtleties of using them to explain the world around you.

Chapter 29 • Reuse Old Envelopes: You can calculate almost anything using only common knowledge–you just need to learn how to put this knowledge to use, and this chapter (which focuses on back-of-the-envelope calculations) is just the thing to help you do it.

Chapter 30 • Use Forecasts with Care: The future is uncertain, but people keep trying to forecast it anyway. Numerous pitfalls await, and without a keen eye for the tricks of this trade, you'll be hard pressed to avoid them.

Chapter 31 • Hear All Sides: In any intellectual dispute, it pays to hear two well-prepared debaters argue their points before drawing any conclusions. Always make such debates fodder for your deliberations, and your decisions will benefit.

PART V • SHOW YOUR STUFF: Once you've done good work, you'll want to present it effectively to readers or listeners. The chapters in this section give insights into making your results "grab" your audience, designing good tables and figures, and using those tables and figures to convey your key points. The section concludes by exploring effective use of the Internet in publishing your analysis.

Chapter 32 • Know Your Audience: The first rule of data presentation is to know your audience and present compelling information in a form your readers can easily grasp. Most analysts forget that other people don't care nearly as much about their results as they themselves do.

Chapter 33 • Document, Document, Document: An astounding number of analysts routinely omit vital data and assumptions from their reports, but you should avoid this pernicious practice. The best analysts document *everything*, giving credit where credit is due, leaving a trail so they can remember, and creating a trail for others to follow. Documentation is also a key step in checking your work, because it forces you to think clearly about your analysis.

Chapter 34 • Let the Tables and Graphs Do the Work: When writing technical reports, create the analysis, tables, and graphs first, then write around them. If the analysis is well thought out, the tables and graphs well designed, and the audience clearly defined, the report should practically write itself.

Chapter 35 • Create Compelling Graphs and Figures: Follow Edward Tufte's rules for graphical excellence, and avoid the most common pitfalls in designing charts and graphs. Your goal should be to give to the reader "the greatest number of ideas in the shortest time with the least ink in the smallest space."[3]

Chapter 36 • Create Good Tables: A well-designed table is a work of art; a sloppy one is worse than useless. Make your tables a resource that your readers will keep as a reference for many years to come.

Chapter 37 • Use Numbers Effectively in Oral Presentations: Even veteran presenters show too many of the wrong numbers. Present only those numbers that support the story you are telling, and focus on that story, NOT on the numbers themselves.

Chapter 38 • Use the Internet: The old ways of publishing are fast being supplanted by web-based approaches. Learn about these new tools and put them to work for you.

CONCLUSIONS • CREATING THE FUTURE: This chapter gives perspective on why we use analysis in the first place. Understanding the world is a prerequisite for making it better!

WHO SHOULD READ THIS BOOK

This book grew out of my experience in training analysts whom I've hired in the past fifteen years. It is written for beginning problem solvers in business, government, consulting, and research professions, and for students of business and public policy. It is also intended for supervisors of such analysts, professors, and entrepreneurs (who may not consider themselves analysts but who need to create analyses to justify their business plans to potential investors). Finally, it covers many topics that journalists who focus on scientific or business topics will find useful.

HOW TO USE THIS BOOK

There is no need to read the chapters in order. Go straight to those that interest you most, but skim the chapters you skip. You just might see something useful there that you did not expect.

Most chapters have "links" to other chapters, with graphical signposts indicating which chapter or major section to investigate for each link (the relevant chapter number appears inside). These signposts look like the link to Chapter 14 that appears in the right margin opposite this line. In the future electronic versions of this book these links will be "live" so that readers can jump to related sections with ease.

14

All Uniform Resource Locators (URLs) discussed in the book are enclosed in triangular brackets, to set them off from the text. They appear as follows: <http://www.lbl.gov>. The brackets and any punctuation marks that precede or follow them are not part of the URL.

All URLs, as well as many key data files, are available in electronic form at <http://www.numbersintoknowledge.com>. If you have questions, comments, or suggestions, you can post them at this site. I'll gladly evaluate them for inclusion in the next edition. I'm particularly interested in examples of large and public analytical blunders by people who should know better, examples of bad or good tables and graphs, and suggestions for how the book can be improved or expanded.

The endnotes contain references, attributions, and detailed information for

the interested reader. The *Further Reading* at the end of the book does not attempt to be comprehensive. Rather, it contains selected sources for each chapter that I regard as most crucial for mastering the material. If I mention a book, I include it in the Further Reading section for the chapter in which I refer to it. At the beginning of the Further Reading section, I also include the "Top Ten" list of my very favorite sources on this topic, which (in my opinion) are the "must read" items that all serious problem solvers should have on their shelves.

Like all other arts, the science of deduction and analysis is one that can only be acquired by long and patient study.

— **SHERLOCK HOLMES**

ACKNOWLEDGMENTS

I would like to thank the following people for their comments and encouragement throughout the process of writing this book: Todd Baldwin, Eric Bergman, Marcela Bilek, Jeanne Briskin, Erik Brynjolfsson, Chris Calwell, Sheryl Carter, Cynthia Chaffee, Eva Cohen, Ken Conca, Laura Driussi, Carol Emert, Marta Feldmesser, Maria Fink, Karina Garbesi, Steve Greenberg, John Harte, Dianne Hawk, Rick Heede, Ephraim Heller, Karen Heller, Chris Koomey, Gregory Koomey, Richard Koomey, Tod Loofbourrow, Amory Lovins, Nathan Martin, Alan Meier, Evan Mills, Marja Mogk, Amey Moot, Mark Morland, Laura Nelson, Laura Newcomer, Andrew Nicholls, Art Rosenfeld, Erika Rosenthal, Marc Ross, Suzanne Samuel, Polly Shaw, Lene Sørensen, and Kim Taylor.

Special thanks to R. Cooper Richey at UBS Warburg and Michele Blazek at AT&T, whose careful reading vastly improved this manuscript.

I would also like to express my gratitude to those friends and colleagues who were most instrumental in teaching me the "tricks of the trade" over the years, including Florentin Krause, Mark Levine, Jim McMahon, Art Rosenfeld, John Harte, John Holdren, Barbara Barkovich, Bill Ahern, and Amory Lovins.

Wylie O'Sullivan and Nan Wishner performed careful copy editing of the manuscript, and Veronica Oliva alerted me to possible issues related to permissions for copyrighted material cited within (though I made all decisions about the use of copyrighted material and permissions for that material). Susanna Tadlock supervised the production and printing of the book (a task for which I owe her special thanks), Sandy Drooker is responsible for its lovely design, and Dave Peattie did a masterful job of typesetting.

Tom Chen's graphics grace the front and back covers and a couple of the chapters. He deserves my thanks for creating art that really works with the book.

Finally, I would like to dedicate this book to my parents, Cynthia Chaffee and Richard Koomey, who always encouraged me to follow my own path. Their advice has proved invaluable more times than I can count.

INTRODUCTION

THE INFORMATION EXPLOSION

In 1815, Thomas Jefferson's library of 6,500 volumes represented an exemplary collection of works by the world's great thinkers. A diligent student who read two books a day, 325 days per year, could peruse the entire collection in ten years. At the time, Jefferson's collection was a large one, and someone who read the library in its entirety could reasonably claim familiarity with the most important knowledge of western civilization.

The same diligent student would have to spend a bit more time to read all the books and manuscripts in the Library of Congress today. At the same rate of 650 books per year, it would take more than 15,000 years to read them all (and that's not even counting the tens of thousands of new books written every year!)[4]

The two centuries since Jefferson, particularly the last half-century, have been extraordinary ones for human knowledge. The pace of change has quickened noticeably, and the rate of information production has increased by many orders of magnitude,[5] driven by rapid progress in science and technology and by vastly improved information transmission, processing, gathering, and storage capabilities.[6] For the first time in human history, information is being created at a rate far faster than humans can assess and use it.

David Shenk, who documents these trends in his book *Data Smog*, reports (among other things) that

> In 1971 the average American was targeted by at least 560 daily advertising messages. Twenty years later that number had risen sixfold, to 3,000 messages per day.
>
> Paper consumption per capita in the United States tripled from 1940 to 1980 (from 200 to 600 pounds) and tripled *again* from 1980 to 1990 (to 1,800 pounds). In the 1980s the use of third-class mail (for sending publications) grew 13 times faster than the population.
>
> Two-thirds of business managers surveyed report tension with col-

leagues, loss of job satisfaction, and strained personal relationships as a result of information overload.[7]

People usually react to such information overload in one of two ways:

- *They keep reading and never shut down the flow of information* in hopes that something new might emerge to help them solve the problem at hand. In this case, people can be paralyzed by too much information.[8]
- *They shut out new information because it's overwhelming.* People afflicted with this reaction are handicapped by too little information. They rely on old knowledge and ideology, and make bad decisions because they close off the flow.

My goal is to help you plot a middle course between these two extremes, giving you tools and tricks to help you face the onslaught of new information with equanimity. The rest of this book will hone your analytical instincts and prepare you to prosper in this information-glutted world.

Our networks are awash in data. A little of it is information. A smidgen of this shows up as knowledge. Combined with ideas, some of that is actually useful. Mix in experience, context, compassion, discipline, humor, tolerance, and humility, and perhaps knowledge becomes wisdom. —**CLIFFORD STOLL**

THINGS TO KNOW

Before we begin our explorations, there are a few preliminary items to work through that can help increase your problem-solving proficiency. You may need to read these chapters a couple of times, but they are worth mastering.

Don't allow yourself to be intimidated by know-it-alls who thrive on bestowing their knowledge on insecure people. Put cotton in your ears and blinders next to your eyes, and trudge ahead with the confidence that whether or not someone else "knows it all" isn't really relevant; the only thing that's relevant is what YOU know and what YOU do.

— **ROBERT RINGER**

CHAPTER

I

BEGINNER'S MIND

In the course of my life I have often had to eat my words, and I must confess I have always found it a wholesome diet.

— **WINSTON CHURCHILL**

Most people attack a new problem by relying heavily on the tools and skills that are most familiar to them. While this approach can work well for problems that are similar to those previously solved, it often fails, and fails miserably, when a new problem is particularly novel or vexing. In these circumstances, it is best to assume nothing and treat the problem as if you have never seen anything like it before.

In martial arts, this sense of looking freshly at something is known as "Beginner's Mind." Beginners to any art don't know what is important and what is irrelevant, so they try to absorb every detail. Experienced martial artists use their experience as a filter to separate the essential from the irrelevant. When that filter mistakenly screens out something essential, then even seasoned masters can err.

The phrase "Beginner's Mind" is a reminder that experience is a two-edged sword. It eliminates unnecessary detail and allows a focus on what are ostensibly the most important elements of a problem. It can also lead you astray when a new problem is sufficiently outside your experience. The art is in knowing when experience is not applicable to the problem at hand.

The ideal is to combine the curiosity and the non-judgmental observation of a beginner with the experience of a senior analyst. Together, these skills can be an immensely powerful combination.

A Japanese Zen master received a university professor who came to inquire about Zen. It was obvious to the master from the start of the conversation that the professor was not so much interested in learning about Zen as he was in impressing the master with his own opinions and knowledge. The master listened patiently and finally suggested they have some tea. The master poured his visitor's cup full and then kept on pouring. The professor watched the cup overflowing until he could no longer restrain himself. "The cup is overfull, no more will go in."

"Like this cup", the master said, "you are full of your own opinions and speculations. How can I show you Zen unless you first empty your cup?" — **BRUCE LEE**

DON'T BE INTIMIDATED

Brenda Ueland, in her classic treatise on creative writing (*If You Want to Write*), explicitly advocates thumbing your nose at "know-it-alls, jeerers, critics, doubters," several times a day. These intimidators murder talent and crush creativity, and you shouldn't let them get away with it.

Most of us have experienced people who confidently toss out statistics to bolster their arguments. Although there are those who speak only when they are sure of their facts, there are many more who express their opinions even when their ideas are ill founded. They do so secure in the knowledge that most people will not see through their game.

In 1974, Robert J. Ringer first published his book titled *Winning Through Intimidation*. In it, he described how he developed his own philosophy for success in his chosen field of real estate sales. The core of this philosophy is what Ringer calls "The Theory of Intimidation," which has wide application to every field of human endeavor. In summary, it states that

The rewards a person obtains are inversely proportional to the degree to which that person is intimidated.[9]

In other words, people do worse in work, life, and love when they are intimidated by others.

To avoid being intimidated, Ringer developed a strong posture. He first created a professional image that would prevent others from intimidating him (for example, he created a lavish "calling card" that was so nicely produced that anyone receiving it would immediately be convinced that Ringer was "somebody"). He then perfected the use of legal tools that would ensure his right to claim rewards for his efforts. Finally, he was so good at his profession that no one could legitimately conclude that his performance was anything but excellent.

Ringer's refusal to be intimidated resulted in spectacular financial rewards relatively quickly. The important lessons to take from his book are that

- many people use intimidation to accomplish their goals even when they are not in the right;
- these people often use (and abuse) analysis to commit these acts of intimidation, and
- someone who knows the right question to ask about how such analysis is created can often deflect acts of intimidation before any damage is done.

Part III With the right preparation, you can beat the intimidators at their own game.

Why do people who know the least know it the loudest?

— **GEORGE CARLIN**

INFORMATION, INTENTION, AND ACTION

In spite of people's differences, there are common threads in the way each of us makes decisions and carries them out. The "Cycle of Action" described in this chapter is not an all-encompassing model, but it can help you understand the role of information in your decision making. It can also help you make better decisions in the future.

Donald Norman, in his book *The Design of Everyday Things*, first introduced the concept of the Cycle of Action (see **Figure 3.1**) and applied it to how people interact with technology. He separates human response to events into Goal Setting, Execution, and Evaluation. Say, for example, that you want to hear some music. Goal setting involves deciding which outcome is most desirable (listening to some music). Execution is the process of influencing external events so that they move more closely to that outcome (you find the compact disc, put it into the player, and turn the player on). Evaluation is the process of observing the resulting external events and determining whether those events are consonant with your goals (if the music starts and you like what you hear, then you've successfully achieved your goal; if not, you need to chose a new disc; if the music doesn't play, you'll need to fix the stereo).

Your decisions are a function both of your past experiences, evaluated after the fact, and your current needs and desires. Your past experience enjoying particular recordings, for example, combined with your current goals of hearing music, will help you decide whether to choose one compact disc or another.

The major steps in the cycle can be summarized as follows:

- Decide what to do, based on your goals;
- Do it (execution);
- Assess the results (evaluation);
- Decide what to do next, and then repeat the process

FIGURE 3.1. The Cycle of Action

SOURCE: Adapted from Norman, *The Design of Everyday Things*

Norman applies the cycle to characterize the choices of individuals, but he could just as well have used it to describe the process by which institutions interact with the external world. Of course, no person or institution follows such a methodical procedure all the time, but you'd be surprised at how much such a simple model explains. Let's delve into more detail, and the value of this model will become obvious soon enough.

There's no identifiable beginning to the cycle (except perhaps the day you were born), but, for convenience, we'll start with Goal Setting.

THE GOAL SETTING PROCESS

You set goals by deciding what you want to happen, based on your values, evaluation of past experiences, and assessment of the current situation. By its very nature, this process cannot be totally objective or value free. If it involves choices, it involves value judgments. To pretend otherwise is both foolhardy and shortsighted.

Peter Hess and Julie Siciliano, in their book *Management: Responsibility for*

Performance, define key characteristics of effective goal setting in the management context, but these criteria apply more generally. The goals must be specific, meaningful to and accepted by the participants, and "realistic yet challenging." Finally, effective goals have a clearly defined deadline and ways of measuring progress toward the goal.[10]

The goal can be as simple as satisfying your hunger or as complicated as influencing the future direction of the computer industry. Once you've decided what outcome you desire, you take action (execute).

THE EXECUTION PROCESS IN DETAIL

The execution part of this cycle can be split into three pieces, as shown in **Figure 3.2**:

- your intention and desire to achieve the goals, which exists because of your human need for closure, fulfillment, status, etc.;
- a list of actions (the "action sequence") that you plan to undertake, which is developed based on your evaluation of the situation, your goals, and your experience; and
- the actual execution of the action sequence, successful completion of which depends on your capabilities.

Developing an intention to act

Execution cannot take place without an intention to act. Before deciding what to do, you first need to decide to do *something*. From this intention flows the next step (a decision about *which* action to take). If you decide not to act, the cycle stops here.

Formulating the action sequence

The action sequence is the set of steps that, if taken in the correct order, will lead to the desired outcome, assuming no unanticipated issues arise. The decision maker develops this sequence based on judgment, experience, and forecasts of likely alternative outcomes for different courses of action.

Accurate data collection and analysis make it more likely that the action

30

FIGURE 3.2. **The execution part of the cycle of action**

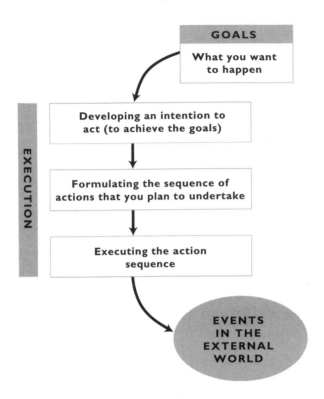

SOURCE: Adapted from Norman, *The Design of Everyday Things*

sequence will be developed in a sensible way in any particular instance. Only rarely does the decision maker have complete knowledge of all relevant factors, so she relies on instinct, experience, ideology, and whatever relevant data have been brought to bear on the topic. Any decision is usually one of dozens to be made on a given day, and there are few such decisions that do not also face time constraints. Poor judgment can result as much from tight time constraints as from data that are compiled inaccurately.

Because all information is uncertain in one way or another, it is important to develop strategies to cope with that uncertainty when formulating the action sequence (for more details, see Morgan and Henrion's book *Uncertainty*). One of the most effective coping strategies involves thinking through alternative scenarios in advance of a particular decision.

26

When evaluating how information was used to develop the action sequence in a particular instance, always explore the interests and training of the people involved. Such an exploration will almost certainly be fruitful. In his book *How to Drive Your Competition Crazy*, Guy Kawasaki points out that "a manager who has been through years of price wars will interpret an opponent's price reduction differently from a Harvard MBA, who's never been closer to a price war than a lecture hall, or a Wharton MBA, who has to formulate an econometric model to buy a pack of gum."[11]

Executing the action sequence

After deciding what to do, you must follow through with action. Good data documentation practice, accurate analysis, and critical thinking can help ensure that the outcome of execution is what you intend. The perceptions of others can be key to success, and a decision maker backed up by credible analysis can influence those perceptions to her advantage.

The execution process itself can also affect goal setting. For example, the very process of taking action can cause you to realize more precisely what you want. It will also give you better information on what it will take to achieve your goal and help you decide whether it's worth continuing to pursue the goal.

EFFECTS ON THE EXTERNAL WORLD

After the action sequence is executed, the decisions affect the external world in some way. For the Cycle of Action, the relevant external world is everything external to the decision process of the individual or institution that might be affected by the action sequence. These effects can be confined to physical ones (i.e., you turn on the stereo after determining the correct action sequence to do it) or to personal/emotional ones (e.g., you decide to play a song on the stereo that reminds you of a long-lost love and brings back your feelings for that person). Usually, though, the effects are both physical and emotional. Often there are many distinct effects of a given action sequence, some of which cannot easily be foreseen. You assess these effects in the evaluation stage of the Cycle of Action.

THE EVALUATION PROCESS IN DETAIL

The Evaluation part of this cycle also can be split into three components, which are illustrated in **Figure 3.3**:

- perceiving the state of the external world, which involves observation, direct measurement, or indirect data collection;
- interpreting the perception, which requires that you apply your knowledge and logic to understanding what you have perceived; and
- evaluating the interpretation and comparing it with your goals. This latter process is the one where value judgments and assumptions most forcefully exert their influence.

Each step in the evaluation process can be influenced by personal biases, which affect the things you choose to measure and how you interpret those measurements. Preconceptions inhibit data collection and cause unusual results to be dismissed too easily. Remember to approach evaluation using "Beginner's Mind," and you'll fall prey to this pitfall less often.

Perceiving the state of the external world

Some aspects of the external world can be measured, but others cannot. Perceiving the external world may include both intuitive and quantifiable knowledge, and each person usually draws on both kinds of knowledge when assessing the current state of the world.

People often measure only those things that are easiest to measure. Like the man looking for his keys under the street lamp (because that's where the light is) even though he lost his key chain down the street, you are likely to be led astray if what you're measuring is only peripherally related to what you care about. If the people charged with collecting the data are too far removed from the actual decision and don't know how the data will ultimately be used, the results can be bad or even disastrous. Guy Kawasaki reports that Sam Walton, the founder of WalMart, did his own research on the competition as he was building his business because *he* was the one who ultimately had to make the strategic decisions. He didn't want data collection choices made by people in his organization who did not understand the big picture.[12]

FIGURE 3.3. The evaluation part of the cycle of action

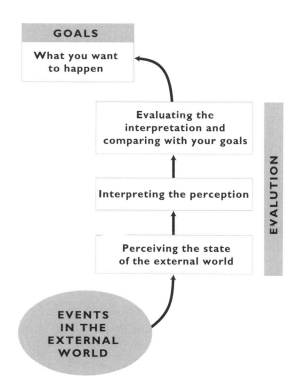

SOURCE: Adapted from Norman, *The Design of Everyday Things*

Kawasaki also describes how businesses can sometimes deliberately skew their competitors' perceptions of the external world. In the late 1960s, Wilson Harrel used his knowledge of Proctor & Gamble's (P&G's) test marketing plans (which were public knowledge) first to make P&G believe that its new cleaning product would be a big success. He temporarily withdrew his own competing product (Formula 409) from Denver, the test market city, which made P&G's test product sell well. P&G then decided to roll out the product nationwide, based on the "success" in Denver. To thwart this launch, Harrell promoted an extremely low price for a large amount of Formula 409 just as P&G launched its national campaign. Harrel's promotion effectively removed a large number of consumers from the national market for cleaning fluid, because the special price applied to enough cleaning fluid to last a typical household for half a year. P&G discontinued its product less than a year later, because of disappointing sales.[13]

Information technology has made it easier to collect data on events we could only dream about tracking twenty years ago. In 1997, Peter Coffee, a columnist at *PC Week* magazine, advocated that businesses collect data on sales of products by day of the week and even by time of day.[14] The benefits of such tracking include improved customer service, inventory management, cash flow tracking, and staff allocation, but it has only been during the past decade that such data collection has become possible on a large scale.

Even with new technology, however, collecting the relevant data is not always straightforward. The ranking of different Internet web sites in terms of numbers of visitors is critical for web advertisers and other businesses, but small differences in measurement methodology can lead to large differences in these rankings. Internet companies are still struggling to define appropriate measurement methodologies, and these difficulties are unlikely to be resolved soon.[15]

Inside an institution, the process of perceiving the state of the external world can be subverted by the complexities of office politics. The busiest decision makers (CEOs, Presidents, Cabinet Secretaries) are often protected from new ideas by their legions of "handlers," from their Chiefs of Staff to their secretaries. Robert Reich discovered this truth when he became Secretary of Labor in 1993:

> My cavernous office is becoming one of those hermetically sealed, germ-free bubbles they place around children born with immune deficiencies. Whatever gets through to me is carefully sanitized. Telephone calls are pre-screened, letters are filtered, memos are reviewed. Those that don't get through are diverted elsewhere.[16]

The Cycle of Action can thus become the Cycle of Inaction when information is diverted from its intended audience. Never forget that, within institutions, the most important decision makers virtually always face some variant of Reich's "bubble." Reich himself tried to puncture the bubble now and then, to the consternation of his staff.

Interpreting the perception

Once data are collected, they almost always require "massaging" into a useful form. This step can involve as little as minor reformatting or as much as detailed calculations and analysis. The choice of how much to manipulate data

One approach...

is as much art as science, and it relies heavily on the instincts of the researcher. This choice is influenced strongly by how particular calculations are to be applied. For analysts who are preparing data for others to use, assessing the needs of those who will use the data is a key step in the data development process. `32`

For the "sales by time of day" example described above, sales data might be viewed for individual products or bundled into groups of products to help discern patterns and trends. The data could also be viewed by hour, by day of the week, by month, or by year.

If the pending decisions are to be made by people other than those collecting and massaging the data (as is often the case), the data must then be summarized in the form of text, tables, and graphs for the decision makers' use (if the information is for your own use, you can make the tables and graphs but skip the text). This step also involves judgments about which data to emphasize. Poor choices here can make irrelevant the most useful data and can lead to erroneous conclusions. Hourly sales data, for example, might be overwhelming if presented en masse—some filtering is required to make these data useful. Such filtering might include focusing on results for one particularly important product or time period, for example.

As Edward Tufte says, bad table and graph design reflects sloppy thinking. At this stage in the cycle, follow the guidelines presented later in this book about designing figures and tables. Don't complicate the process of "interpreting the `35` `36` perception" by creating tables and graphs that obscure the data and confuse your audience.

Evaluating the interpretation

The evaluation of data involves drawing conclusions, developing assertions, and deciding which of these assertions to emphasize. These assertions can take several forms, but the choice of which ones to make from a given set of data is at the discretion of the analyst (that means YOU!) or the people presenting information to the decision maker. Even among the set of assertions that can be logically supported by a particular set of data, there is no single correct choice here.

At this point in the cycle, the analyst must make explicit assumptions about relevant unknowns (based on experience, values, instinct, and ideology), or simply ignore them. Usually, these issues are noted and put aside for future work. Ignoring these unknowns can lead to serious analytical errors, as can making bad assumptions about them. Unfortunately, there's no way to avoid this risk, but you can insulate yourself from it by creating credible scenarios.

The evaluation at this stage can be affected by power relationships between the analyst and the source of the data. Jeffrey Laderman, writing in *Business Week* in 1998, describes how ratings of stock analysts on particular companies have changed over the past twenty years. **Figure 3.4** summarizes his results. Although "Sell" recommendations were tendered roughly one-quarter of the time in 1983, such recommendations had virtually disappeared by 1998. The percentage of "Buy" recommendations grew from about one-quarter in 1983 to two-thirds by 1998. Laderman writes that "by the end of the 1980s, Buys outnumbered Sells four to one, and by the early 1990s, eight to one."

In part, this shift was the result of the ongoing bull market in the U.S., but it was also partly caused by the increasingly cozy relationships between investment banks and the companies their stock analysts covered. Banks whose analysts dispensed "Sell" recommendations were shut out of competing for lucrative public stock offerings by those companies. Their analysts were also excluded from company briefings. Company pressure undeniably affected (and probably continues to affect) stock analysts' evaluation of the interpretation when assessing company stocks.

Once the evaluation stage is completed (you've compared the outcome of execution with your goals), then the cycle begins anew. If things didn't work out as you had hoped, you should start formulating an action sequence that might lead to the desired result next time.

FIGURE 3.4. Percentage of "buy," "sell," and "hold" stock recommendations by U.S. analysts in 1983 and 1998

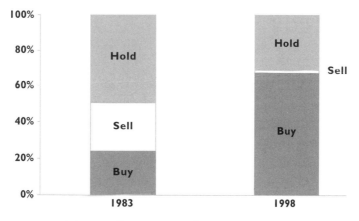

SOURCE: Laderman, Jeffrey M. 1998. "Wall Street's Spin Game: Stock Analysts Often Have a Hidden Agenda." In *Business Week*. October 5. pp. 148-156.

THE BOTTOM LINE

The Cycle of Action reminds us that making decisions and taking action rely on choices about which data to ignore and which to address explicitly. It also illustrates that unknown and unknowable data, as well as experience, intuition, and assumptions, affect *every* choice. No decision is immune from these facts, and you would do well to keep them in mind.

Time constraints can influence all the steps in the Cycle of Action. Rarely do decision makers have enough time to evaluate all the options, so they rely on judgment and ideology to help them assess a situation. Execution of plans also occurs in a time-constrained environment because other people and institutions are making changes that can negate the beneficial effects of your own changes, creating pressure to act quickly.

Refer to the Cycle of Action whenever your choices make events go awry. It can be applied to most decisions in your life, from those about what clothing to wear to those about which college to attend. If you think systematically about how you use information to make decisions, your decision making will improve in the future.

FIGURE 3.5. The complete cycle of action in detail

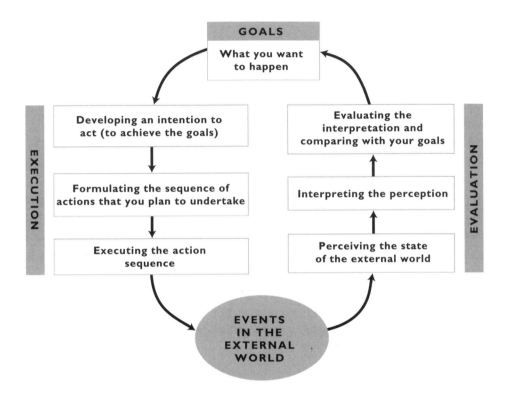

SOURCE: Adapted from Norman, *The Design of Everyday Things*

CONCLUSIONS

The collection and use of data are human endeavors and, as such, are subject to all the vagaries of individual and institutional behavior. At any point in the Cycle of Action, human frailties can change the course of events. This realization should not inspire fear or resignation. Rather, you should refer back to the complete Cycle of Action in **Figure 3.5** whenever you want to improve your use of information or to explain how a decision-making process broke down in a particular instance. Identify which parts of this cycle were the result of your own choices, those of others, or external events. Such a systematic assessment can help you use information more skillfully and make better decisions in the future

EXERCISE

Think about three decisions you made in the past week. Did your decision process follow the Cycle of Action in all cases? Where did your biases affect the outcome? How did you decide what your goals were? Did you have enough time to interpret and evaluate your perceptions of the external world before making the decision?

It is not the critic who counts, not the man who points out how the strong man stumbled, or where the doer of deeds could have done them better. The credit belongs to the man who is actually in the arena; whose face is marred by dust and sweat and blood; who strives valiantly; who errs and comes short again and again; who knows the great enthusiasms, the great devotions, and spends himself in a worthy cause; who, at the best, knows in the end the triumph of high achievement; and who, at worst, if he fails, at least fails while daring greatly, so that his place shall never be with those cold and timid souls who know neither victory nor defeat.

— **THEODORE ROOSEVELT**

PEER REVIEW AND SCIENTIFIC DISCOVERY

Science is a human process that advances in fits and starts. The end result of this process is generally something you can count on, but the intermediate steps to this result do not advance in a linear progression as science textbooks might lead you to believe (see Kuhn's *Structure of Scientific Revolutions*). Understanding the process by which scientific knowledge is created can help you better assess its implications and relevance to your own analysis.

Most people (especially nonscientists) don't recognize the large role that intuition and instinct play in scientific discovery, particularly in the process of developing hypotheses. A scientist's desires for recognition and status, enthusiasm for solving a problem, and attitudes towards authority are all motivations rooted in human needs and emotions.

These human inclinations all affect the progress of science, but the end result of the scientific process is one that is rational to the core. Science has an internal coherence and predictive power that is unique among human endeavors. Historian of science Gerald Holton identifies the "apparent contradiction between the seemingly illogical nature of actual personal discovery and the logical nature of well-developed scientific concepts."[17]

This stark contrast between the messy business of scientific discovery and the end result, which is the most accurate description possible of how the physical world operates given current knowledge, comes about in part because of a process known as "peer review." It is important for nonscientists to understand how this process works, because accurate peer review, more than any other part of the scientific process, determines whether a particular set of research results is credible.

THE IMPORTANCE OF PEER REVIEW

The formal process of peer review is most often conducted for articles submitted to scholarly ("peer-reviewed") journals or for research proposals submitted to funding agencies (such as the National Science Foundation). In journal reviews, the author submits the article to the editor, who then distributes the article to between one and four reviewers. The reviews are sometimes "blind" so that the reviewers don't know the names of the authors and their affiliations (it is however often possible for knowledgeable reviewers to infer the identities of the authors from a paper's citations and the authors' approach in presenting the results). More often, the authors' names are known to the reviewer. Only rarely is the *reviewer's* name revealed to the authors when comments are delivered. Both the number of reviewers and whether the reviews are blind are at the discretion of the journal editor. Reviews of scientific project proposals are usually more formalized than are journal reviews as befits a process upon which research funding depends directly.

The purpose of peer review is to judge whether the work is based on principles and judgments that represent the current scientific consensus. When such a consensus exists in a particular field, it reflects the paradigm accepted by scientists in that field and gives practitioners a standard set of tools with which to evaluate new ideas (see Kuhn's *Structure of Scientific Revolutions* for more details). When a paradigm is challenged by anomalous observations and experiments, a crisis ensues. The crisis continues until a new paradigm emerges that encompasses the anomalous results. During periods of crisis (or in non-scientific fields) peer review is generally less reliable because the foundations upon which the reviewers judge research results in the field are being questioned in fundamental ways.

Adequate documentation is a pillar of the peer review process. Without it, scientists could not verify or reproduce results, and scientific progress would grind to a halt. The best scientists are fanatical about documentation, and you should be too. **33**

Peer review is not a guarantee of accuracy. In fact, papers containing major scientific blunders have been published in peer-reviewed journals.[18] However, the formal peer review process makes it more likely that a paper will avoid major flaws. Be skeptical of "research" results that are not peer reviewed. There are some "scientists" who announce their results to the media but do not pub-

FIGURE **4.1. A comical view of how human frailty can affect the course of science**[19]

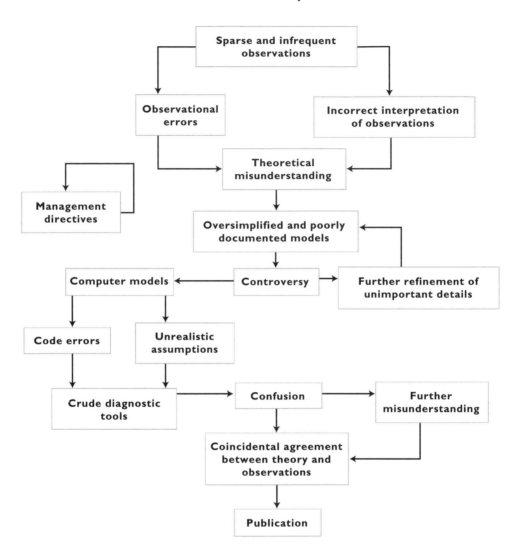

lish in peer-reviewed journals. Usually, such announcements are funded by particular organizations with an ax to grind and have little to do with science.

The "cold fusion" episode in 1989 is one example where the announcement of a "discovery" in a press conference but not in the refereed literature led to a media circus. Two scientists at the University of Utah (Martin Fleischmann and Stan Pons) claimed that they had created a room-temperature process that generated significantly more energy than it took to run the experiment, and that they had detected evidence that a nuclear fusion reaction was responsible. Had

the press been more skeptical of a research result announced only in a press conference, its pronouncements might have been more cautious. Instead, the claim of limitless cheap energy resounded loudly around the world, only to fall to the ground with a sickening "thud" when experiments by other scientists failed to corroborate the results.

How can you tell if a scientific study has been peer reviewed? **Table 4.1** lists some important peer-reviewed journals for key scientific fields. If the study is published in one of these journals, it's passed the first hurdle of basic scientific credibility. For a long (but still probably incomplete) list of peer-reviewed journals, check out <http://adswww.harvard.edu/abs_doc/refereed.html>. For a ranking of the importance of different journals in virtually all scientific fields, see the *Science Citation Index*, put out annually by the Institute for Scientific Information or ISI <http://www.isinet.com>. These rankings are based on the number of citations to papers in a given journal by the scientists in that field, so they are a relatively objective measure of journal quality. In its electronic database, ISI's web site also has a large searchable list of journals, as well as other related products.

SCIENTIFIC PROOF

There are two kinds of proof accepted in scientific research. The first, based on deductive logic, relies on assumptions and general principles to derive implications and predictions of specific events.[20] If I leave my bicycle outside and it rains, I deduce from my experience and the laws of chemistry that parts of the bicycle will rust. Deductive inferences rely on a set of initial assumptions ("axioms"), which are analyzed using rules of logic. If the initial assumptions are correct, the rules of logic are followed, and no logical contradictions arise in the analysis, then the conclusions must be correct.

The second kind of proof—inductive logic, on which many inferences in science and other forms of human endeavors are based—relies on compilations of specific instances to infer general laws from the specifics (see Hughes, *Critical Thinking*). If I leave my bicycle unlocked outside for a hundred days and it is never stolen, I could use inductive logic to infer that it will not be stolen tomorrow, but I cannot be sure. I can only say that it is *unlikely* to be stolen, based on

TABLE 4.1. Some important peer-reviewed scientific journals

Field	Journal
Basic Science/Medicine	Science Nature Cell Nature Medicine
Clinical Medicine	New England Journal of Medicine Journal of the American Medical Association Annals of Internal Medicine
Physics	Physical Review Physical Review Letters Nature
Chemistry	Science Accounts of Chemical Research Journal of the American Chemical Society
Physical Chemistry	Journal of Physical Chemistry Journal of Chemical Physics Chemical Physics Letters
Biology	Science Nature
Geology/Geophysics	Journal of Geophysical Research — Solid Earth Geophysics Applied Geophysics
Global climate	Global Biogeochemical Cycles Tellus Biogeochemistry
Ecology	Ecology Ecological Applications
General Environmental Science	Environmental Science and Technology Atmospheric Environment
Hydrology	Water Resources Research Journal of Contaminant Hydrology
Combustion Technologies	Combustion and Flame Combustion Science and Technology
Electrochemical Technologies	Industrial Engineering & Chemical Research Journal of Power Sources Journal of Electrochemical Society
Physical Chemistry Technologies	Applied Surface Science Journal of Non-Crystalline Solids Applied Spectroscopy
Biological Technologies	Methods in Enzymology Biochemical Pharmacology American Journal of Physiology

my extrapolation of past experience. Inductive proof is less certain than deductive proof (in rare instances it can be *as* certain, but never more so).

The *Encyclopedia of Philosophy* defines inductive argument to include "all cases of nondemonstrative argument, in which the truth of the premises, while not entailing the truth of the conclusion, purports to be a good reason for belief in it." [21] In plain English, an inductive proof shows that the conclusion is highly probable but not demonstrably true in the same way that a correct deductive proof is. Induction is the link between mathematics, deductive logic, and our experience of the physical world.

Deduction by itself is impotent because it depends only on assumptions and its own internal logic. Without induction to enrich the set of data and assumptions upon which deductions can be based, the world of deduction would be a spare one, indeed. When inductive logic was first formalized, the goal of its devotees was to show that its results were demonstrably *true*, in the same way conclusions based on deductive logic were true. According to the *Encyclopedia of Philosophy*, "not until the end of the 19th Century did a more modest conception of inductive argument and scientific method, directed toward acquiring probability rather than certainty, begin to prevail." [22]

The limitations of induction are offset in part by the peer review process. Outside reviewers uncover errors of fact, logic, and omission, and report them to the authors and the journal. Obvious errors are weeded out quickly while more fundamental flaws may take years, decades, or even centuries to be uncovered. Eventually, though, even these errors will be discovered as scientists attempt to build on previous work. Serious flaws will result in logical inconsistencies (discovered through deduction) that will inevitably surface and be corrected.

THE SCIENTIFIC OUTLOOK

The physicist Alan Sokal points out that science is predicated on two key attitudes:

- being willing to accept what you find; and
- being willing to discover that you are wrong.

Human nature being what it is, these two attitudes are rare. Next time you find

yourself resisting a new idea, take a deep breath and try to see the other point of view. Ask yourself why this idea might make you uncomfortable (often it's because it clashes with your ideology). If you can step back from a situation in this way, you will have achieved what I like to call the true "scientific outlook."

CONCLUSIONS

Science and technology are a critical part of modern life. You owe it to yourself to understand at least the basics of how scientific knowledge is created and used.

Remember always that science is a human endeavor. Learn about the process so you can better appreciate how to interpret scientific findings. Finally, always give more weight to peer-reviewed research than to results announced solely in the media.

EXERCISE

Find a newspaper story summarizing a scientific study that interests you. Find out where the author of the study works. Who funded the research? What's the name of the journal in which the research was originally published? Is it a refereed journal? Send for the report and compare it to the newspaper article's summary of its conclusions. Is the newspaper summary accurate?

The supreme task ... is to arrive at those universal elementary laws from which the cosmos can be built up by pure deduction. There is no logical path to these laws; only intuition, resting on sympathetic understanding of experience, can reach them. — **ALBERT EINSTEIN**

BE PREPARED

Few people are so brilliant that they need never prepare, and even they would do far better with a little work beforehand. Preparation is one of the foundations for success in any field. The six short chapters in this section describe important preparation that any good analyst should consider. Prepare well, and you'll never regret it.

If I had eight hours to cut down a tree, I'd spend six hours sharpening my ax. — **ABRAHAM LINCOLN**

CHAPTER

5

EXPLORE YOUR IDEOLOGY

Anyone who states that ideology has no effect on decisions is either lying or deluded. Everyone has an ideology whether explicit or not. Ideology provides a simplified model of the world that reflects our values, biases, and experiences. It helps people make decisions in the face of imperfect knowledge.

Anton Flew's *Dictionary of Philosophy* defines ideology as "any system of ideas and norms directing political and social action," indicating the key role it plays in mediating between ideas and choices. Ideologies are often explicitly political (libertarian, conservative, liberal) but may also relate to preferences about material things.

Even technical subjects can be affected by ideology. In late summer 1997, certain hackers with a libertarian ideology actually protected a recently released version of the Apple Macintosh operating system from being illegally copied (even booby trapping or removing copies from public newsgroups) because they wanted to strike a blow against Microsoft, the dominant company in the computer industry.

Computer users each see themselves in ways that reflect their ideologies. Apple Macintosh users, for example, often see themselves as going against the conventional wisdom–resisting the crowd mentality by choosing a computer they regard as technically superior (Apple feeds that perception in its advertising, associating, for example, Albert Einstein's unconventional brilliance with its products. See <http://www.apple.com/thinkdifferent>). Microsoft Windows users, on the other hand, stereotypically see themselves as realists who are adopting the standard because they perceive its complete dominance to be inevitable, because they believe it is cheaper, because they value the greater availability of software for Windows, or because they are forced to use it at work.

Ideology is one connection between the world of ideas and the world of choices. It impinges upon the Cycle of Action in the assessment of unknown or unknowable aspects of events and in the collection and interpretation of data

for events that can be measured. Ideology guides the assumptions and value choices you make in the course of any analysis, and it can be particularly helpful (or particularly misleading) in situations where your knowledge is limited.

There's no guarantee that your ideology, which has been developed in an iterative process between your predispositions and your experience, will work well in the future even if it's worked well in the past. That's why ideologies that maintain their relevance over time must allow for adaptation and evolution when confronted with new data and experience.

My own ideology, based on reason, pragmatism, environmental consciousness, and social responsibility, has not changed much during the past two decades. One core part of this ideology is, in the physicist Alan Sokal's words, "a respect for evidence and logic, and for the incessant confrontation of theories with the real world."[23] My ideology is constantly challenged by friends who find my rock-solid belief in reason distasteful, but thus far I have not been convinced to modify that part of my ideology in a significant way. Over the years, I have changed positions on particular issues, but my core belief in logic and evidence has never wavered.

The ideology held by an individual *can* change radically over the course of a lifetime. Eldridge Cleaver, one of the original Black Panthers in the 1960s, became a conservative Republican as he grew older. Such shifts can be triggered by a watershed event, or can happen gradually as a person's experience affects his or her outlook.

It is essential to think through your ideology so that it will be consistent. You can check its *external* consistency by examining how its assumptions and conclusions stack up against facts you know for sure. Sometimes there is no way to check external consistency, and we say in this case that the assumptions behind the ideology are not falsifiable, testable, or provable. We can always test *internal* consistency, however. If there are internal inconsistencies, we know there's something wrong.

To help think about your ideology, ask yourself these questions:[24]

- Where do you get your information?
- Whom do you trust? Why do you trust them?
- With whose opinions do you normally agree? Do you agree with all their opinions? Why do you agree? Do you believe in their values or do you trust their judgment?

- To whom do you give the benefit of the doubt? Why?
- Are there events that outrage you? Why do they make you mad?
- What motivates you to work hard? From where did those motivations come?
- Could one part of your ideology lead you to take actions that would be inconsistent with another part of your ideology?

Ideology is the filter through which you view the world. By exploring your own body of beliefs, you can more reliably make choices that are consistent with your values, and you will also be better prepared to use numbers responsibly and effectively in your analyses, writing, and presentations.

EXERCISE

After you've asked yourself the questions above about your own ideology and distilled the answers into a coherent credo, try to answer the same questions for a few people whose ideology is quite different from your own (or ask them the questions directly!). This may help you understand them better, which is the first step toward true communication.

Only the supremely wise and the ignorant do not alter.

— CONFUCIUS

GET ORGANIZED

Many people garner high scores on standardized tests but are unable to apply themselves in a way that gives results. *Raw intelligence* is the ability to think quickly and accurately; *useful intelligence* is raw intelligence as enhanced by hard work, common sense, instinct, experience, discipline, and organization. A person of moderate raw intelligence who possesses high useful intelligence will virtually always be more productive than a brilliant person who lacks the personality traits needed to make that brilliance useful in the world.

Organization is the single most important factor in creating useful intelligence, and it is a *learned* skill. By consciously becoming more organized, you can substantially increase your effectiveness in virtually any endeavor.

Julie Morgenstern, in her book *Organizing from the Inside Out*, gives sound and practical advice for getting organized. By following her systematic approach for assessing your goals and analyzing your current state of disorder, you can begin to get control of the chaos. After you've used Morgenstern's tricks and tips for assessing and addressing the situation, you can take the following additional steps to achieve effective personal organization:

- *Control information flows*: Decide which information streams are relevant and which are detrimental to your professional effectiveness. Cancel magazine and newspaper subscriptions except those you use regularly. Limit your use of the World Wide Web to get news updates—with few exceptions, checking more than once a day is a waste of time and effort. Set aside specific times to answer mail and email (there's usually no need to respond to emails as they arrive). Recycle junk mail without opening it.
- *Buy a paper or electronic organizer*: Track to-do tasks, appointments, addresses, phone numbers, and notes in an orderly way. Get rid of those myriad slips of paper that confuse and confound.

 A trick I recently discovered to keep track more effectively of to-do tasks

is to create a blank page with different types of actions (like "call," "do," "email," "fax," "investigate," "find," "visit") and to write down your tasks under the appropriate category (see **Figure 6.1**). This approach immediately creates a task-oriented hierarchy that helps you prioritize. I keep separate pages like this for work and home to-do tasks. If things are really busy, you can even keep separate pages for different projects.

- *Clean your office*: Throw out extraneous stuff and store rarely used books. Make frequently used materials easily accessible. Keep only current work on your desk and put other materials in drawers and filing cabinets. Identify those items that are important to you and make them accessible. Morgenstern says "keep what you use and what you love. Toss the rest." Spend fifteen minutes at the end of each day to tidy up your office, so you can start fresh in the morning.

- *Develop a filing system*: Filing is for retrieval, *not* storage. Create a filing system that helps you (or others) retrieve stored information in a timely fashion. If you can't find an article or electronic file, it's worse than useless because you waste time looking for it. Be selective about the items you file. `7`

- *Take time to reflect*: Make sure you give yourself at least a half-hour per day of solid thinking time, with no interruptions. Unplug the phone and turn off the radio. `21`

- *Prioritize tasks and focus on the essential*: Too often, our days are consumed by "urgent tasks that are not important." For at least part of the day, focus on the "important tasks that are not urgent."[25] `10`

- *Document, document, document*: Documentation creates a trail for you and others to follow after the fact. It is essential to any intellectual work, but is often neglected. By making it a priority, you ensure that your intellectual contribution will remain important and useful for years to come. `33`

You have control over each of these steps. Do them, and you'll make the most of every opportunity. Avoid them, and you'll never be as effective as you could be. Even the most creative people, who often view organization as inimical to their personal style, would benefit greatly from making it part of their daily routine.

EXERCISE

Think of three important goals in your life that you were unable to achieve because you were disorganized. Remember those missed opportunities when you feel overwhelmed by the clutter, and use those examples to motivate your cleanup.

EXERCISE

Copy the example to-do list (**Figure 6.1**) and give it a try in your daily life. Alternatively, create a similar list that more closely reflects your needs.

In human endeavor, chance favors the prepared mind.
— **LOUIS PASTEUR**

FIGURE 6.1. An Example Task List

Call	Date Due	Do	Date Due
Give/Deliver	Date Due	Get	Date Due
Email/Search on Web	Date Due	Find	Date Due
Fax	Date Due	Investigate	Date Due
Mail	Date Due	Delegate	Date Due
Talk to/Visit	Date Due	Arrange	Date Due

ESTABLISH A FILING SYSTEM

A key part of being organized is creating a filing system that helps you retrieve stored information in a timely fashion. Such a system is critical for both printed documents and (increasingly) computer files. If you can't find a file, it doesn't matter if you have it or not, and looking for it distracts from other more vital tasks.

Be selective about the items you file. For years I clipped articles from newspapers on subjects in which I maintained an active interest, thinking that someday I might write papers on those topics. In most cases I never did, and all that effort was for naught. There was one example where the files came in handy for a research project, but I can only think of that one. I stopped clipping most newspaper articles years ago, but I still save a few articles and reports that I find particularly valuable for current projects.

The futurist Peter Schwartz also gave up an elaborate filing system because he never used it. His approach is one to emulate: "I concentrate on educating myself; on passing information through my mind so it affects my outlook; on tuning my attention as if it were an instrument." He recognizes that with this approach he may not be able to draw upon previously compiled materials, but that can be a benefit in disguise: "Sometimes I must go back and re-create all the research I did several years before. But that, in itself, is valuable, because in the fields I care about, the facts have changed since I last went to look for them." Moreover, he saves time by only filing those items most crucial to his work.[26]

Paper should pass through your hands only once on the way to the file, to the recycling bin, or to the outbox. You may never achieve this ideal in practice, but strive for it and you will save time and be more effective.

Never clip anything you can easily find later, especially if you can search for it electronically. As more newspaper articles and reports have become available on the Internet, my need for hard copy storage has decreased. Current events are well documented on the World Wide Web, and you need only call up one of

the many competent search engines to get a nice listing of recent articles on particular topics. Many newspapers even offer CD-ROMS or web retrieval of older articles, which makes clipping articles even less worthwhile. Such electronic retrieval sure beats clipping and filing articles! If you need to create your own database of electronic clippings, there are many capable sheet scanners that will quickly and efficiently put these clippings on your computer.

All reports that my group creates at Lawrence Berkeley National Laboratory are put on the World Wide Web, in a file format (Portable Document Format, or PDF) that will allow all types of computers to read it (Adobe distributes their Acrobat Reader Program for free to anyone who wants to download it, at <http://www.adobe.com/products/acrobat/readstep.html>). Others can then download our reports without the time and expense of having to send the hard copies. I also don't have to worry about where the copies of the printed reports are—I can just call them up in my web browser and view them. Often this is easier than hunting around in the filing cabinet for an old report. To supplement this electronic storage, I organized by subject area three-ring binders that contain our most recent reports, for ready reference.

Order files by active projects and by major subject areas. Also have file drawers for people, institutions, and for administrative issues. My "People" drawer holds files for particular individuals who work in my field (it is usually easier for me to remember who worked on a particular report than to remember the subject heading where I filed that report—having a file for individuals gets around that problem). The "Institutions" drawer is analogous to the People drawer. Create an archive for old projects that is in one of the less accessible filing cabinets, and put the most widely used "active" files near your workspace.

Frequent travelers should allocate one part of a file drawer to "Travel," and create one file folder per trip, with destinations and dates clearly marked on each tab. Order them chronologically and save them for a long time. Years later, you may need to know details of a certain trip, for tax or other purposes.

A key trick to making any filing system work is to go through your files periodically to familiarize yourself with where things are. This occasional sorting will give you the opportunity to discard archaic materials and refile items that belong somewhere else. No filing system is ever finished. Periodic sorting helps keep your system current and useful.

The process of developing a filing system will teach you something about your work. To create such a system, you will need to analyze and categorize the kinds

of work that you do, and you will surely learn something from this effort. You will also need to think carefully about the kind of information you most often retrieve from your files. If others must also be able to access your filing system, you might adopt a slightly different organization than if you are the only user.

In a World Wide Web browser like Netscape Navigator or Microsoft Explorer, it is easy to record "bookmarks" for sites that you frequently visit. It's important to organize these well, so create a structure that parallels your hard-copy filing system. These bookmarks are crucial to fast information retrieval.

Computerized reference databases like EndNote can make any filing system more effective. Each report is one record in the database, and it is a trivial matter to add a sentence in the notes section of each reference record that indicates the file drawer in which the hard copy is stored. These databases can be searched electronically, which can be a godsend when you're looking for a crucial report.

To preserve access to your stored articles and papers, DON'T lend them out to others unless this lending is recorded in a loose leaf binder devoted to this chore. I routinely reach for a document on my shelves at work, only to find that someone has borrowed it. Such a written record can help you track down missing reports when this happens.

Most people borrow reports and then forget they have them. You should insist that someone borrowing a report copy the relevant pages (obeying copyright laws, of course) and return it to you promptly. This insistence will save you precious hours in the long run.

EXERCISE

List major topic areas for your current work. Compare these topics to the headings of your filing system. How well do they match up? Do the same comparison for your computer files, electronic folders, and bookmarks saved in your web browser. Is it time to reorganize things?

There's nothing remarkable about it. All one has to do is hit the right keys at the right time and the instrument plays itself.

— **JOHANN SEBASTIAN BACH**

8

BUILD A TOOLBOX

A problem solver needs a toolbox as much as a machinist needs a lathe. My toolbox is the set of tricks and techniques that I have mastered and can use to solve problems. Each person's toolbox is different and is a product of his or her education and experience.

No toolbox is ever complete because every problem has its own unexpected subtleties; the toolbox that worked for past problems may need to be updated to face new challenges. The learning process never ends.

Part I

LANGUAGE

Each field has its own language. You must at least develop working familiarity with basic concepts in a given field before attempting to decipher a problem. If you are delving into a field for the first time, obtain a textbook and begin your reading there. Textbooks conveniently summarize the core information in a given field and usually provide a nice glossary and lists of further references. Another good place to begin is the list of frequently asked questions (FAQs) on a particular topic that is available on the Internet for virtually any issue you can name <http://www.faqs.org>. Talk to people you trust who know something about the topic because they can often point you to the best reference sources **23** and save you time.

CALCULATIONS

An important part of any analyst's toolbox is the ability to create simple back-of-the-envelope calculations at a moment's notice, using approximations and **29** memorized data. Such calculations are a quick way to check the quantitative assertions of others and are invaluable in both business and science.

Darrell Huff's book, *The Complete How to Figure It*, is a beautiful compilation of hundreds of ways to calculate numbers important to daily life. Huff is the fellow who blessed the world with the book *How to Lie with Statistics*. Some of the major headings in the book, just to give you a flavor, are as follows:

> Lifetime money strategy
> Some personal things (like life expectancy)
> Interest and saving
> Investing
> Other people's money
> House planning
> Building things (like additions to your house)
> Your car
> Travel
> Conversions

The book is chock-full of specific examples (more than 350!) that you can copy when doing your own calculations. I recommend adding it to your toolbox.

Another nice source of math techniques is a book written by Clifford Swartz called *Used Math*. It is more advanced than Huff's book, and summarizes the basics of calculus and other essential mathematical tools, mainly for students of science. If you find yourself needing to refresh your knowledge of more advanced math, this book is a great place to start.

The best textbook I've found on practical applications for mathematics is called *For All Practical Purposes*. It covers five major areas: management science, statistics (the science of data), coding information, social choice and decision making, and size and shape. Most mathematics texts shy away from real-world applications, but this one breaks the mold.

For an accessible and entertaining book about statistics, try Larry Gonick and Woollcott Smith's *The Cartoon Guide to Statistics*. Don't let its title fool you. It's a solid introduction to statistical concepts and methods.

CALCULATIONAL COMPUTER TOOLS

To investigate any particular software package, I recommend that you browse in your local bookstore. Writing computer books has become a cottage indus-

try, and within weeks of the release of new software a book appears on the shelves to assist you with its use.

Spreadsheets and databases are the two most common computer tools used for calculations. The first spreadsheet was VisiCalc, which was created in the late 1970s and was a great innovation compared to the programming languages that had been commonly used for business applications before that. I use spreadsheets every day. They are fine for simple analyses and powerful enough for moderately complex modeling tasks. They are also handy for graphing results and creating formatted tables for reports.

Databases are useful when you have large amounts of data that all have the same format. For example, if you have a mailing list of three hundred names, with addresses, phone numbers, and email addresses, a database is the right tool for organizing it. A database will allow you to export the data in any format. You may want names and addresses in one case, but names and email addresses in another. Once entered into the database, you can extract them any way you want. Spreadsheets can often serve as simple databases as well.

A more advanced set of tools is available for statistical analysis of data. These software packages can be quite expensive and are generally for expert use only. They are worth investigating if you need the horsepower and have the knowledge to use them effectively. They can help you divine relationships between factors that are related in a complex way. Many of these complex analysis functions are also available in spreadsheets, which can be useful for such analysis if the data sets aren't too large and the number of variables to analyze is small.

28

EXERCISE

Browse through the computer help section at the bookstore, and read about software you don't currently use. Think about a problem that you're wrestling with now, and see if there's a way to apply some of these unfamiliar tools to solve it.

Applying computer technology is simply finding the right wrench to pound in the correct screw. — **ANONYMOUS**

PUT FACTS AT YOUR FINGERTIPS

My college buddies used to object to my tactics in our numerous discussions about public affairs because I would be the first to pull out a reference book to check their sometimes wild assertions.[27] While your friends may also think it unfair, your discussions will be far more fruitful.

GENERAL

The single most important document for a U.S.-based researcher in virtually any quantitative field is the *Statistical Abstract of the U.S.* It is available every year from the U.S. Superintendent of Documents and other sources for around $40. You can also buy a CD-ROM with all the Statistical Abstract's tables in electronic form or you can download those data for free from the web. It's a real bargain, containing about 1,500 tables summarizing everything from air pollutant emissions to population, average temperatures to voter registration, and hunting licenses to federal budget outlays. The CD-ROM version is especially nice because it allows easy access to tables in spreadsheet format, giving you a head start on your analysis.

The *Statistical Abstract* is the best source for items (such as population and **27** economic activity) needed to normalize any data you get from other sources (for example, absolute growth in cement production over time is not easily understandable until you normalize it on a per capita or per dollar of GDP basis). It also contains the latest information on inflation over time, for use in converting any current dollar statistics to constant (inflation-adjusted) dollars. The data are **33** taken from a variety of sources and are carefully documented. The footnotes to the tables will give you leads on sources that you never knew existed. The *Statistical Abstract* is a great place to start any investigation, and you should have a recent electronic or hard copy version available at all times. All

libraries will have the latest one. To order *The Statistical Abstract* or to download tables from it go to <http://www.census.gov/statab/www/>.

A companion to the *Statistical Abstract* is a good almanac. Although some data in these two sources overlap, the almanac contains many useful items that *The Statistical Abstract* doesn't have, such as zip codes, postal rates, tax rates, historical events, sociopolitical facts about states and countries, and other information that you didn't know you really needed (but you really do). There are many almanacs; the one I have is *The World Almanac and Book of Facts*. *Time Magazine, The New York Times,* and *The Wall Street Journal* also publish almanacs, some of which are more specialized than others. An almanac of some kind will be available in any bookstore or library, or you can access the Information Please Almanac on the web at <http://www.infoplease.com/>.

Many newspapers and magazines offer searchable archives of past issues on the web or on CD-ROM. Using such archives makes initial research on new topics relatively easy. *The New York Times,* for example, allows web readers to search the past 150+ years of archived articles and to see the results of the search in summary form <http://www.nytimes.com>. Downloading a single article costs $2.95 (as of this writing), which is charged to the user's credit card. Many public libraries offer access to similar CD-ROMS and on-line databases, so check your local library before buying.

Another excellent source of data is a book by Mary Blocksma called *Reading the Numbers*. This book summarizes the meaning of numbers we encounter in everyday life, like the codes found on the side of tires or the conventions used in describing fluorescent light bulbs. It is unfortunately out of print but can be obtained at used bookstores, <http://www.alibris.com>, and some libraries.

A similar book that is more readily available is Andrea Sutcliffe's *Numbers: How Many, How Long, How Far, How Much*. The variety of data contained in this book is truly astounding. In one section there are estimates of safe food storage times in the refrigerator and freezer, and in another there is a list of the number of different types of instruments in a typical symphony orchestra. The major sections include money, food and nutrition, health, travel, daily life, transportation, time, weather, people and places, the universe, conversion tables, and everyday math. The book contains more than 600 pages of data, and it's well worth having on your shelf.

The World Resources Institute <http://www.wri.org>, in *World Resources*

1998-99, and Lester Brown et al. in *Vital Signs 1999*, present essential information related to resource consumption and environmental effects around the world. These data include pollutant emissions, freshwater resources, land use, population and health, food and agriculture, wildlife and habitat, energy, atmosphere and climate, and policies and institutions. The data in *Vital Signs* are also available electronically from the Worldwatch Institute <http://www.worldwatch.org>.

WEB/INTERNET DATA

More data appear on the Internet every day. It is an impossible task to report the latest developments in a book like this one (things change too fast), but there are a few key sites of which you should be aware (All URLs and changes to them can be found at <http://www.numbersintoknowledge.com>):

- *The Library of Congress site* <www.loc.gov> is an invaluable source. You can search many parts of the library's databases with little effort. Many local libraries have also placed their catalogs on the Internet, which can save you a trip to the library next time you want to do research.

- *The U.S. Bureau of the Census site* <http://www.census.gov> is beautifully designed and is a treasure trove of social, demographic, and economic data. You can search by key word, by place, by clicking on maps, and in many other ways. Most of the data on this site are free or available at nominal cost. This site also houses the on-line version of the Statistical Abstract of the U.S. <http://www.census.gov/statab/www/>.

- *The National Technical Information Service* (NTIS) is still, as of this writing, the official source of many U.S. government reports. NTIS sells almost three million titles, and you can search on its nicely designed web site <http://www.ntis.gov> to find the one you want.

- *The U.S. Department of Energy's Energy Information Administration site* <http://www.eia.doe.gov> is the right place to find energy and some environmental data for the U.S. The site also has limited information on certain international topics, such as nuclear power or energy use in other

countries. Most of the data on this site are free or available at nominal cost.

- *The U.S. Environmental Protection Agency's site* <http://www.epa.gov> contains huge quantities of environmental data, as well as access to the EPA's reports.

- *The Fed Stats site* is a treasure trove for U.S. economic and other data <http://www.fedstats.gov>. It allows easy access via the World Wide Web to data from more than 50 federal agencies.

- *The Economagic web site,* which has both U.S. and international economic data, is located at <http:// www.economagic.com>. This site claims to offer access to 200,000 data files (many of them for free), with special services for subscribers (fees for subscribers range from small to moderate depending on the level of service).

- *The National Center for Health Statistics site* <http://www.cdc.gov/nchs> contains data and reports on U.S. health that can be downloaded for free.

- *The U.S. Geological Survey* produces a superb on-line atlas for the United States, which can be found at <http://www.nationalatlas.gov>. This site allows the user to map standard political boundaries, identify the locations of airports and dams, and overlay the routes of roads and railroads. It also contains maps of Super-fund sites, air pollution emissions, mining operations, industrial facilities, and selected data from the U.S. Census.

- *Frequently Asked Questions (FAQs)* exist on the Internet for almost any topic <http://www.faqs.org>. Because these FAQs are compiled by experts in each field, they are a great place to begin your inquiries. Use Internet search engines to find the FAQs that will help you most.

- There are hundreds of *newsgroups* on everything from Star Trek to statistical analysis. By posting an email to the relevant newsgroup, you can often get answers to questions from people all over the globe in a matter of hours or days. You can generally access these newsgroups using your web browser (consult your Internet service provider for details, or go to <http://groups.google.com>).

There's a real art to searching for information on the Internet. Each different search engine brings to you only a subset of the information actually available, so you'll need to use several to be certain you've done a thorough search. For

example, when I searched for the words "information overload" on the Alta Vista search engine, it returned about 1,000 documents. When I used Infoseek, it returned just under 5,000. Not all of the items on the Alta Vista list were in the Infoseek list, and vice versa. Skill is also involved in narrowing down your search so that you have at most dozens of documents to sort through, instead of the thousands that these engines commonly produce for you. Search engines try to rank their results by relevance, but the algorithms for ranking the relevance of documents also differ.

In recent years, Google <http://www.google.com> has become the dominant search engine. A convenient site where you can access other commonly used search engines is <http://www.freeality.com/metay.htm>. Also check out <http://dir.yahoo.com/Computers_and_Internet/Internet/World_Wide_Web/Searching_the_Web> or <http://www.searchenginewatch.com>, which give search strategies and details on specific search engines. Each search engine uses different algorithms and syntax for defining a search. It's worth mastering your favorite search engine's tricks so that you can find information quickly and efficiently.

CONTEXT

Sometimes in presenting the results of analysis, you'll want to know the intellectual history behind a certain concept. For example, if you are giving a talk about energy policy to a general audience, you might want to open with a brief summary of how the concept of energy was developed and how our modern conception differs from that common in the 1800s. In this circumstance, two essential works are *The Encyclopedia of Philosophy* and *The Oxford English Dictionary*. The former is helpful in tracing the genesis of a particular concept through the ages while the latter tracks the origins of particular words. Each of these sources can be useful in writing a speech or a paper and are both found in most libraries.

An alternative to the conventional *Encyclopedia of Philosophy* is *The Stanford Encyclopedia of Philosophy*, which is a web-based encyclopedia of comparable scope to the hard-copy version, on line at <http://plato.stanford.edu>. It is con-

stantly being updated and evaluated by its editorial board (they call it a dynamic encyclopedia). The creators of this new tool felt that the old way of creating an encyclopedia was too slow to capture the rapid developments in philosophical thinking (particularly developments related to scientific concepts). For a more general on-line encyclopedia, go to <http://www.wikipedia.org>.

A good book of quotations is essential to liven up your writing and presentations, and a standard dictionary and thesaurus will make your writing more precise. These are available in myriad incarnations in any bookstore. If you have easy access to the web, a large collection of quotations can be found at <http://www.cyber-nation.com/victory/quotations/quotes_menu.html>. You can search Roget's thesaurus at <http://humanities.uchicago.edu/forms_unrest/ROGET.html> or the Merriam Webster thesaurus at <http://www.m-w.com/thesaurus.htm>.

For many academic topics, excellent summaries of the current state of research are contained in *Annual Reviews*. These reviews are published for 27 different fields, including political science, public health, energy and the environment, and a wide variety of physical and biological sciences. They are often the best place to start when you're trying to learn about a new field.

CONVERSION FACTORS

For physical constants and conversions, there's nothing better than *The Handbook of Chemistry and Physics*. It comes out every year, but, because physical laws and constants rarely change, even an old version of this work is handy, so you should pick one up at a used bookstore. Virtually all libraries have a copy of this book on the shelf.

There are certain facts that I use frequently. For example, I often need to convert English units to metric units and vice versa. To help this process along, I memorize the most commonly used conversions and keep the rest on a one-page data sheet for ready reference. I recommend compiling such a sheet for your own use that contains the conversions most useful to your work. The hour or so of work needed to compile this page will be repaid many times over in the convenience of not having to dig deeply into obscure reference books whenever you need to do a calculation. See **Figure 9.1** for an example data sheet.

EXERCISE

Compile a data sheet for your own use after thinking for a week or two about the numbers you use most frequently. Pass it around to your co-workers to get their suggestions. Print copies for them when you're done, to share the wealth.

Data, Data, Data! I can't make bricks without clay!

— **SHERLOCK HOLMES**

FIGURE 9.1. **Example Data Sheet**

Physical Constants and Conversions (underlined values are exact)

Reprinted with permission from J. P. Holdren, January 2000.

1 ton (t) = 1,000 kg = 2204.6 lb = 1.102 short ton * 1 lb = 453.6 g * 1 oz = 28.35 g
1 atomic mass unit = 1.6606e-24 g * 1 grain = 6.480e-2 g * 1 carat (metric) = 0.20 g
1 m = 3.281 ft = 39.37 in = 1e6 micron = 1e10 angstrom = 1e15 fermi * 1 in = 2.54 cm
1 km = 0.6214 mi = 0.5400 naut mi * 1 ft = 0.3048 m * 1 parsec = 3.0857e16 m
1 m2 = 10.76 ft2 = 1e28 barn * 1 km2 = 100 hectare (ha) = 247.1 acre = 0.3861 mi2
1 mi2 = 2.590 km2 = 640 acres * 1 ft2 = 929.03 cm2 * 1 in2 = 6.452 cm2
1 m3 = 999.97 liter = 35.31 ft3 = 264.2 gal * 1 ac-ft = 1233 m3 * 1 km3 = 0.2399 mi3
(1 liter = 1 kg of H2O at 4 degrees C; density of H2O at this T is 0.99997 g/cm3.)
1 cord = 128 ft3 = 3.625 m3 * 1 bushel = 3.524e-2 m3 * 1 Imp. gal = 4.546e-3 m3
1 m/s = 3.281 ft/s = 2.237 mph = 1.944 knots * 1 mph = 0.869 knots = 1.467 ft/s
1 cfs = 0.0283 m3/s = 449 gpm = 724 ac-ft/yr * 1 m3/s = 35.3 cfs = 25.6e3 ac-ft/yr

1 J = 1e7 erg = 1 W-s = 6.241e18 eV = 0.7376 ft-lb * 1 Btu(IT) = 1055.1 J (Internat
Table Btu&cal = 1.00067x thermo Btu&cal) * 1 cal(th) = 4.184 J, 1 cal(IT) = 4.187 J
1 kWh = 3.6 MJ = 3412 Btu(IT) = 859.8 kcal(IT) = 1.341 hp-hr * 1 hp = 550 ft-lb/sec
1 t TNT = 1e9 cal = 4.184e9 J (TNT explosive energy = 900-1100 cal/g)
1 a.m.u. converted to energy = 1.492e-10 J * 1 quad = 1e15 Btu = 1.0551e18 J
1 TW = 31.536 EJ/yr = 29.89 quads/yr = 14.12 Mbbl(US oil)/day = 1.076 Gt(UN coal)/yr

1 t coal equivalent (tce) [UN,WrldBnk,Shell] = 7,000 Mcal(IT) (low heat) = 29.31 GJ
1 oil equiv (toe) [BP,Shell] = 10,000 Mcal(IT) (low ht) = 41.87 GJ —> 5.7 GJ/bbl
1 bbl crude oil [USDOE] = 5.8e6 Btu (IT, high heat) = 6.12 GJ —> 44.9 GJ/ton
1 t crude oil [UN] = 1.454 tce = 42.6 GJ * 1 t crude oil = 7.33 bbl = 1.166 m3
1 bbl petroleum products [USDOE] = 5.48 MBtu = 5.78 GJ * 1 bbl gasoline = 5.54 GJ
1 tce [BP] = 0.667 toe = 27.9 GJ = 26.5 MBtu * 1 t utility coal [DOE] = 24.76 GJ
1,000 m3 natural gas [BP,WorldBnk] = 9,000 Mcal(IT) (low ht) = 37.68 GJ; [Shell] =
9,500 Mcal(IT) (high heat) = 39.77 GJ; [UN] = 1.332 tce = 39.0 GJ * 1 t = 1,084 m3
1,000 ft3 nat'l gas [USDOE] = 1.028 MBtu -> 38.3 MJ/m3 * 1,000 kWhe [BP] = 0.25 toe
= 10.47 GJ (low); [DOE] = 10.47 Mbtu (11.0 GJ) hydro, 11.02 Mbtu (11.6 GJ) nucl (high)
1 t U3O8 @ 0.5% utilization = 336e3 GJ —> 30.3 GWh(e) if converted at 32.5%
10 m3 water/sec x 10 m head = 0.98 MW * 1 t air-dried wood (1.39 m3) or dung = 15 GJ
1 t charcoal = 29.3 GJ (produced @ 25% effic. in dirt kiln, 50% in masonry kiln)
1 t crop wastes (20% moisture) = 13 GJ * 1,000 liter kerosene = 37.6 GJ
C02 emissions 25 kgC/GJ coal, 20 kgC/GJ oil, 15 kgC/GJ natural gas

1 newton (N) = 1e5 dynes = 0.2248 lb(force) * 1 pascal (Pa) = 1 N/m2 = 1e-5 bar
1 atm = 101,325 Pa = 10.33 t(force)/m2 = 14.696 lb/in2 = 760 mm Hg (0 C) = 760 torr
1 curie (Ci) = 3.7e10 dis/s = 3.7e10 becquerel (Bq) * 1 rad = 100 erg/g = 0.01 gray
1 gray (Gy) = 1 J/kg * 1 roentgen (R) —> 2.58e-4 coulomb(+ or -)/kg dry air
roentgen-equivalent-man (rem) = rads x quality factor * 1 sievert (Sv) = 100 rem

EARTH: R = 6357/6378(po/eq) km; A = 510.1e6 km2 (oc 361.3, ld 148.8, ice-free ld 133
Asia 50.0, Africa 30.3, N. America 22.4, S. America 17.8, Oceania 8.5, Europe 4.9);
M = 5.97e21 t (oc 1.4e18, ice 2.9e16, atm 5.2e15, lks/str 1.3e14, dry biom 1.8e12);
E @ top of atm = 172,000 TW (surf 88,000, evap 41,000, wind 2,000, GPP 200, NPP 100)
Highest elev = 8848 m; avg elevation = 840 m; deepest ocean = 11,000 m.
1 degree of latitude = 111 km; 1 minute = 1.85 km. Solar constant = 1368 W/m2.

ATM: @ sea lvl 1.225 kg/m3, 288 K, mol wt 28.96, 2.55e19 molecules/cm3, sound speed
331.4 m/s; percent by vol N2 78.09%, O2 20.95%, A 0.93%; ppm—CO2 340, Ne 18, He 5.2,
CH4 2.9, Kr 1.1, N2O 0.5, CO 0.1, O3 0.01; atm water vapor & clouds 1.3e13 t.
Heat cap of air = 1 J/g-K. Tropopause: h = 11 km, p = 0.23 atm, T = 217 K. Wet-air
adiabatic lapse rate = 6.5 K/km, dry = 10 K/km. Viscosity = 1.8e-4 dyne-sec/cm2.

H2O: heat fusn = 79.7 cal/g; heat cap = 1 cal/g-K; heat vapn = 539.6 cal/g (100 C),
584 cal/g (21 C), 597 cal/g (0 C). Ocn vol = 1.35e9 km3; avg depth = 3.7 km; mixed
layer depth = 80 m; sea wtr density (15 C) = 1.026 g/cm3, salinity = 35,000 ppm (Na
10,500, Cl 19,000, S 885, C 28, N 15, Li 0.17, P 0.07, Fe 0.01, U 0.003, Au 0.00001)
Viscosity = 0.01 dyne-sec/cm2.

g = 9.807 m/s2 * R = 8.3144 J/mole-K * N = 6.022e23/mole * c = 2.998e10 cm/s
k = 1.3806e-23 J/K * s = 5.670e-8 W/m2-K4 * K = C + 273.15 * F = 1.8xC + 32
AMU = 1.661e-27 kg = 931.5 MeV * e = 1.602e-19 coulomb * h = 6.626e-34 J-sec

VALUE YOUR TIME

Mr. Casals, you are 95 and the greatest cellist that ever lived," a young reporter commented. "Why do you still practice six hours a day?" Casals replied, "Because I think I'm making progress."

Americans always seem to rush around. There's never enough time to do everything you want to do, see everyone you want to see, or travel everywhere you want to go. In part, this busyness is an inherent part of modern-day life, but it is also partly self-imposed.

Your time is your life, and it is precious. If someone is wasting it, they are stealing it from you, and YOU are allowing that theft. Many people don't think about this equation, and they allow people and institutions to monopolize their time and sap their energy without commensurate rewards. You must manage your time to reflect your priorities (no one else will do this for you).

The most articulate statement of this principle is found in *Your Money or Your Life*. In this book, Joe Dominguez and Vicki Robin describe a set of simple techniques to help people internalize this equation and make it part of their daily decision making. They encourage people to evaluate their spending by calculating how many hours of labor ("life energy") will need to be spent for each purchase. The value of money becomes more real after you ask yourself two key questions:

- "Did I receive fulfillment, satisfaction, and value [from this purchase] in proportion to life energy spent?"
- "Is the expenditure of life energy on this purchase in alignment with my values and stated life purpose?"

"No, Thursday's out. How about never—is never good for you?"

One hundred dollars has an abstract quality. It is made more real when comparing various purchases that can be made for that amount. It becomes most real when put in terms of the four or eight or twenty hours of labor needed to earn the money to buy this new toy you want.

This line of argument dates back to Thoreau, who, in *Walden,* compared the cost of train fare to the hours of time needed to earn the money to pay for it, and concluded that he would arrive in less time by walking. It has implications for your choices of where to work and how to spend your personal time.

Stephen Covey, in his book *The Seven Habits of Highly Effective People,* describes how most people allow urgent items to dominate their lives, but the most effective people focus significant attention on *"the important items that are not urgent."*[28] Be selective about the endeavors on which you spend time. Prioritize. Choose jobs with the greatest likelihood for success and the highest potential payoff. Working at a dead-end job can sap time and energy from a more valuable use of your time.

Think carefully about time spent with others: Does it add to your life? Does

DILBERT reprinted by permission of United Feature Syndicate, Inc.

it make you happy? If not, then do something about it. Don't let anyone steal your life!

Avoid going to committee meetings except where the agenda is well focused and the person chairing the meeting will keep things on track. Most meetings wander into irrelevance, and most people who call meetings have no respect for the time and money wasted. To avoid this problem, one executive prohibited meetings of greater than ten minutes in length.

Avoid large meetings except when they are absolutely essential to improving the productivity of those attending. I knew my meetings were valuable when my staff of 12 actually requested that our biweekly group meetings last for a full hour instead of the 35 minutes we typically took, to allow more in-depth discussion of each person's current research. Of course, some meetings are held to foster teamwork or for political reasons, and these may not be so easily avoided. If you must attend, bring reading or other work so that, during the inevitable slow times in the meeting, you will be able to accomplish something.

You must protect your most productive time from interruptions. Each of us works best at particular times of day. I find that my best writing occurs in early afternoon and early evening. My best calculations are done in the late morning. A very small number of hours account for the bulk of my writing and research output, and if I don't write during those times, I'm vastly less productive.

Find your most productive times and guard them jealously. Unplug the phone. Go to the library. Take control of those times! Your life can be stolen by constant interruptions just as well as it can by people wasting large blocks of time in meetings.

In their classic book, *Peopleware*, Tom DeMarco and Timothy Lister write about how software companies can increase the productivity of their employ-

ees. Software designers are among the most creative people anywhere, and their productivity and accuracy are dependent in part on a workplace that isn't so noisy that it interrupts their work. *Peopleware* describes how programmers who found their workplace "acceptably quiet were *one-third more likely* to deliver zero defect work."[29]

Protect your work area from stray noise. If your office is close to others who are distracting you, perhaps you can negotiate with them. If that's not possible, then work in the library or at home. Don't force yourself to work under noisy conditions: you'll be wasting your time and your life.

If you are feeling unproductive, do something that takes little mental energy or effort but will yield dividends later (like sorting your files, reading mail, or answering phone calls). Even "down time" can be useful if you match the task to your mental energy level.

Finally, value other people's time just like you value your own. Don't schedule needless meetings. Rarely hold long meetings. Be prepared for all meetings. No one ever complains when meetings finish early.

EXERCISE

Keep a log tracking your time for two or three typical days, writing down the topics you worked on. Record all interruptions as well as the duration of uninterrupted time. Are you making the most of your productive hours? Are you surprised by how many interruptions there are?

Furious activity is no substitute for understanding.
— H. H. WILLIAMS

ASSESS THEIR ANALYSIS

Any time you're exploring work others have done, you'll need to know what to look for. Your own systematic thinking can uncover logical blunders that the less critical observer will overlook. Hugh and Lillian Lieber (*The Education of T. C. MITS*) give many examples of apparently obvious conclusions that are erroneous, as does Darrell Huff in *How to Lie With Statistics*. William Hughes, in his book titled *Critical Thinking*, explores in a more general way correct and incorrect modes of argument.

By investigating carefully and thoughtfully, you will be able to evaluate other people's analysis as a matter of course and thus contribute to your understanding of the task at hand. The following chapters give some advice on how to make these explorations productive ones.

Just because I use a study to refute another study does not mean my study is right. It just means I believe it. Caveat Emptor.

— **CYNTHIA CROSSEN**

THE POWER OF CRITICAL THINKING

Years ago, Dr. Norman Vincent Peale became famous for proclaiming the power of positive thinking. His claim was that each of us has control of how we feel about our lives, and, by using that control, we can create a better future for ourselves.

For assessing the analysis of others, *critical thinking* is at least as important as positive thinking. Critical thinking implies the ability to assess the true meaning of arguments and the skill to determine whether those arguments are credible and compelling. William Hughes, in his excellent book *Critical Thinking*, defines some of the key terms for understanding this important skill:

> When we express an inference in words, we do so by means of statements. A *statement* is a sentence (i.e. a set of words) that is used to make a claim that is capable of being true or false.... An *argument* is a set of statements that claims that one or more of those statements, called the *premises*, support another of them, called the *conclusion*. Thus, every argument claims that its premises support its conclusion.[30]

A *logically strong argument*, according to Hughes, is one "whose premises, if true, support its conclusion." A *sound argument* is "a logically strong argument whose premises are true." Critical thinking is the way to identify and create sound arguments, which is the main goal of all analysts.

Hughes describes the key steps a critical thinker uses to assess arguments and decide whether they are sound.[31] I summarize his steps below:

- *"Identify the main conclusion"*: Rephrase the main conclusion in your own words, to be sure you understand it. Think about the context in which the argument is being made. Finally, if the argument is not stated clearly, interpret the argument in the way that is most charitable to your opponent's views (Hughes calls this "the principle of charity").

- *"Identify the premises"*: Rephrase the premises in your own words. List the unstated premises and assumptions. Discard distracting and irrelevant examples that are not really premises. Don't forget that context is important, and that the principle of charity should guide your interpretations at this stage as well.

- *Identify the logical structure of the argument and determine how the premises support the conclusion*: Determine whether one premise depends on another one, or, alternatively, if premises are independent of each other. Determine whether the argument contains sub-arguments that support the ultimate conclusion. Hughes presents a way of describing the structure of arguments, known as a tree diagram, that can help with this step.

- *Ask whether you'd be justified in accepting the premises*: In most cases, a strict proof of the premises is impossible, so we must be satisfied with the less stringent standard of "acceptability." Hughes suggests two criteria for evaluating acceptability:

 — If the statement is common knowledge, we should regard it as acceptable, unless the context requires a higher standard of proof.

 — If the statement is not common knowledge, we should ask for, or be prepared to offer, the evidence upon which it is based, and accept it only if the evidence meets the appropriate standard, for example, personal experience, appeal to a recognized authority, or strict proof.

Hughes also points out that "common knowledge" is often wrong, so use care when relying on any measure of acceptability less than a true proof.

Key questions to ask when determining acceptability include: Are the premises common knowledge? What is the context of the argument? Who is making the argument? Are they experts? Do they have a vested interest in the outcome? Are the premises backed up by credible research published in peer-reviewed journals? Are the premises based on consistent comparisons?

- *Ask whether the premises are relevant*: Many arguments are peppered with premises that are irrelevant to the conclusion but appear to support it. Usually the previous steps will make it obvious which premises don't apply, but it's worth asking the question again at this stage just to be sure.

- *Ask whether the premises "adequately" support the conclusions*: Certain statements lend some support to the conclusion but are not sufficiently compelling to make the argument adequate (appeals to authority often fall into this trap). Inadequacy is a key weak point of many arguments. This

step is mainly an exercise in common sense and in figuring out whether a particular argument accounts for all of the many possibilities.

- *Look for missing arguments:* Omitting a valid comparison is one commonly used way to make an argument seem stronger, so try to find these omissions in other people's arguments. People usually look for the best-case comparisons to support their own position and ignore comparisons that might weaken their case.

- Finally, *"Look for counter arguments"*: Try to find a sound argument that leads to a conclusion contradictory to the one reached by the argument being evaluated. If you find such a counter argument, you know that something's wrong, because two sound arguments cannot contradict each other (unless one is not truly sound).

Careful critical thinking is at the root of all good analysis. When these steps become second nature, you will have mastered its essence. The many subtleties of assessing arguments will always keep you on your toes though. There's no end to the ways that arguments can be flawed.

EXERCISE

In a newspaper, find a letter to the editor on a topic you care about. Identify the conclusion and the main premises. Assess each premise, determining whether each is acceptable, relevant, and adequate to support the conclusion. How easy was it to carry out Hughes's process?

The fascinating impressiveness of rigorous mathematical analysis, with its atmosphere of precision and elegance, should not blind us to the defects of the premises that condition the whole process. There is perhaps no beguilement more insidious and dangerous than an elaborate and elegant mathematical process built upon unfortified premises.

— **T. C. CHAMBERLAIN**

12

NUMBERS AREN'T EVERYTHING

Don't confuse things that are merely countable with those that really count.[32] Some analysts systematically ignore things that can't be turned into a number and assume that because these things can't be measured, they are unimportant. These analysts also generally assume the converse, that things that can be measured must be important. Both mistakes are common, especially among the quantitatively inclined, but these are errors to which you need not succumb.

Four categories of things affect decisions in the world:

- Some things can be measured directly and with confidence before decisions have to be made. The amount of money in your bank account at the time of your last statement falls into this category.

- Some (perhaps most) things are not directly measurable but can be estimated accurately enough in time to influence decisions. You cannot know exactly how much money you will have in your bank account in a month, but you can estimate it in a crude way by making assumptions about your expenditures, based on an extrapolation of past spending trends and planned expenses.

- Some things cannot be measured, estimated, or predicted accurately in time to influence decisions. Some reasonable estimate might be possible for such factors, but time is often limited. For example, time constraints may prevent you from doing a complete analysis of your assets before rushing to put a bid down on a house.

- Some things cannot in principle be measured, estimated, or predicted under any circumstances. Sometimes there are physical reasons for this inability (e.g., the Heisenberg uncertainty principle limits the accuracy of measurements of the velocity and position of subatomic particles) and at other times we cannot measure or predict because things that really matter to people's lives (such as loyalty, love, honesty, and integrity) are sim-

20

What's the value of the earth?

ply not quantifiable in any meaningful way. It is not possible to estimate the true value of a warm summer's day or a species saved from extinction except in terms that have nothing at all to do with numbers.

Events in the last two categories are typically treated in the Evaluation stage of the Cycle of Action (I call these unknown or unknowable aspects of events). To analyze such situations, we must make assumptions and value judgments, using our ideology and instinct as guides to fill in the facts that we can't obtain with measurement or estimation. **3 19 5**

Items in the last category are commonly ignored but often are of the most importance to human life. Because many things simply can't be quantified, we can only insist that decision factors that defy quantification be listed and described in as complete a form as possible, given current knowledge. Then, at least, decision makers are aware of the other factors that may relate to the decision at hand and can weigh these factors based on values, instincts, experience, ideology, and knowledge. If decision makers persist in ignoring the intangibles, **5** then the situation has moved beyond the realm of analysis, and an explicitly political debate about the value of those intangibles must then ensue.

EXERCISE

Think again about the three recent decisions that you explored in the Cycle of Action chapter. Which of the data affecting those decisions could be measured directly? Which could be estimated and which could not? How did you treat the intangibles that could not be quantified? What assumptions did you make to treat the unknowables?

To laugh often and much; to win the respect of intelligent people and the affection of children; to earn the appreciation of honest critics and endure the betrayal of false friends; to appreciate beauty; to find the best in others; to leave the world a bit better, whether by a healthy child, a garden patch or a redeemed social condition; to know even one life has breathed easier because you have lived. This is to have succeeded. — **BESSIE A. STANLEY**

ALL NUMBERS ARE NOT CREATED EQUAL

A common fallacy to which even sophisticated analysts sometimes fall prey is to look for immutable laws of human behavior, in much the same way as the physicist discovers physical laws through experiment. Such generalizations about human and economic systems fail because these systems are adaptable in ways that physical systems are not.

Physicists conduct replicable experiments to uncover fixed natural laws. A scientist measuring the speed of light in a vacuum, for example, would find the velocity the same in the U.S. or Tahiti. If another scientist conducted a similarly accurate experiment in Russia, that test would produce the same result.

On the other hand, relationships between cause and effect for individuals and human institutions are dependent on institutional, social, and economic context. Furthermore, these relationships change over time. A market researcher attempting to predict consumer acceptance for a new toothpaste would find that a market test in San Francisco would likely yield quite different results than in Tahiti even though the speed of light remains the same in both places. In addition, if the same experiment had been conducted in the 1950s, the results would have presumably varied wildly from those of the current day.

In the mid-1970s and early 1980s, conventional wisdom held that modern societies could not reduce energy use without also reducing gross domestic product (GDP) and harming their economies. For example, Gordon Corey, vice-chairman of Chicago's Commonwealth Edison, stated in 1981 that "there is an unbreakable tie between economic prosperity and energy use." Similarly, the Chase Manhattan Bank stated, in its 1976 *Energy Report,* that

> there is no documented evidence that indicates the long-lasting, consistent relationship between energy use and GDP will change in the future. There

Sherman's Lagoon reprinted with special permission of King Features Syndicate.

is no sound, proven basis for believing a billion dollars of GDP can be generated with less energy in the future.[33]

Believers in an unbreakable link between energy use and GDP assigned the immutability of a physical law to this historical relationship but found their belief shattered by events. From 1973 to 1986, U.S. primary energy consumption stayed flat, but GDP rose 35% in real (inflation-adjusted) terms. These believers had forgotten that people and institutions can adapt to new realities, and historically derived relationships (like the apparent link between energy use and GDP that held up for more than two decades in the post-World-War-II period) can become invalid when events (like the 1973 oil embargo) overtake them.

Most economic computer models embody historical experience through relationships that are derived statistically and then use those relationships to forecast the future. These models are often used to assess the potential effects of proposed changes in government policy or business strategy. The models embody history but cannot give an accurate picture of a world in which the fundamental relationships upon which they depend are in flux. If the statistically derived relationships embedded in such a model are the very ones that would be affected by choices or events, then those relationships must be modified in the analysis; otherwise, the results are suspect.

For example, some economists trying to assess the costs of reducing carbon emissions assume that the only way to reduce these emissions is to impose a large carbon tax, which would raise the price of coal, oil, and natural gas in a

way that has no historical precedent. They further compound this error by using models that embody historical relationships to do long-term forecasts without altering those relationships to reflect the unprecedented nature and size of these taxes or the likely effects of other policy choices (see DeCanio 2003).

Creating a world with vastly lower carbon emissions presupposes massive behavioral and institutional changes that render past relationships between energy use and economic activity largely irrelevant (just like what happened after 1973). It also presupposes alterations in government policies. It is laughable to use computer models based on historical relationships for one-hundred-year forecasts in the face of such changes. Instead, scenario analysis is a more `26` appropriate tool for exploring the key relationships and how the world might evolve if those relationships change.

If you are confronted by results of computer-generated forecasts, always ask `30` whether the analysts changed the key historical relationships to reflect the possibility that those relationships will be affected by institutional choices or other developments. If not, then you've identified a key failing of the analysis, and the results are suspect.

I have found that many physical scientists, computer modelers, and economists are susceptible to this mistake. I am not certain why forecasters from these disciplines fall into this trap, but I have noticed it often. It may be what my social-science-oriented friends call "physics envy," or it may be that most analyses are conducted in a mechanical way without significant reflection. In any case, once forewarned, you need not let them get away with it.

It is fashionable today to assume that any figures about the future are better than none. To produce figures about the unknown, the current method is to make a guess about something or other—called an "assumption"—and to derive an estimate from it by subtle calculation. The estimate is then presented as the result of scientific reasoning, something far superior to mere guesswork. This is a pernicious practice that can only lead to the most colossal planning errors, because it offers a bogus answer where, in fact, an entrepreneurial judgment is required. — **E.F. SCHUMACHER**

CHAPTER

14

QUESTION AUTHORITY

Sacred cows make the best hamburger. — **MARK TWAIN**

"Question Authority" is one of the favorite T-shirt and bumper sticker slogans in Berkeley, California. It dates back to the 1960s when students engaged in well-publicized defiance of authority figures in the campus administration and the state government. Although some aspects of these emotion-charged times are best left in the past, it is time to dust off "Question Authority" for the 21st Century.

There are, of course, excellent reasons for following authority in many instances. People can become authority figures by virtue of hard work, unique experience, special insight, and long training. In some situations, such as military maneuvers, obedience to authority is essential to completing missions successfully. Following authority in a business setting can also be important and can allow people to act in a concerted and collective way to achieve the common goal of increased profit for a firm.

However, most situations in everyday life are not as dangerous as military maneuvers. There are two main reasons why questioning authority is often a healthy response:

- Even brilliant people affiliated with well-respected institutions are not always right. History is littered with examples of wrong-headed conventional wisdom and misjudgments by otherwise outstanding individuals. Recall what a Yale University management professor said about Fred Smith's paper proposing reliable overnight delivery service: "The concept is interesting and well formed, but in order to earn better than a "C," the idea must be feasible." Smith went on to found Federal Express Corporation,

which now does billions of dollars in business carrying overnight packages every year.

- Many salespeople and marketing professionals use deference to authority to gain advantage and influence consumer behavior. Robert Cialdini, in his book *Influence*, documents dozens of amusing and sometimes disturbing examples of how these "compliance professionals" use authority figures and pseudo-authority figures to sell products. Cialdini describes one famous example where the actor Robert Young, who used to play Marcus Welby, M.D. on TV, advises people to buy Sanka coffee for its health benefits: Cialdini concludes "objectively, it doesn't make sense to be swayed by the comments of a man we know to be just an actor who used to play a doctor. But, as a practical matter, that man moved the Sanka."

So long as there is time to reflect and consider the dictates of authority, it's healthy to question it. Hughes, in his book *Critical Thinking*, proposes five criteria to use when assessing the assertions of an ostensible authority figure.[34] I summarize them below:

- *"The authority must be identified"*: Appeals to anonymous authorities are immediately suspect. You can't assess the strength of an appeal to authority without knowing who the authority is.
- *"The authority must be generally recognized by the experts in the field"*: At a minimum, the authority must possess an academic degree, a professional license, or an institutional position that indicates acceptance by the relevant expert community.
- *"The particular matter in support of which an authority is cited must lie within his or her field of expertise"*: The authority must be a real expert on the topic at hand, but it is common for experts in one field to express opinions about another field, and be taken seriously whether their opinions are well founded or not (the Robert Young example described above is a case in point).
- *"The field must be one in which there is genuine knowledge"*: Hughes defines genuine knowledge as "a systematically ordered body of facts and principles that are objective in nature." Beware of appeals to authority in fields that don't fall under this definition.
- *"There should be a consensus among the experts in the field regarding the particular matter in support of which the authority is cited"*: Even in

physics, there are issues that remain a subject of controversy. Appeals to authority are only relevant on topics that are not controversial.

Unless these criteria are satisfied, an appeal to authority is not *adequate* to support a particular argument and should probably be disregarded.

Cialdini also suggests asking one more question, namely "how truthful can we expect the expert to be here?" By asking whether the authority figure (or the institution funding her) stands to benefit from the advice she is supplying, we gain insight into the trustworthiness of that advice. If experts have a vested interest in one or more potential outcomes of your decision, their advice should be discounted and you should seek corroborating information. The advice may nonetheless be sound, but it is worth special care in this case.

If the expert's interests coincide with yours or the expert is demonstrably impartial, then more weight should be ascribed to his or her opinions. Even in this case, however, it is best to seek corroborating evidence because experts can be mistaken.

One recent example where the interests of the researchers and their funding institutions led to widespread (and well-justified) public skepticism is the furor over research on the health effects of smoking. For years, the tobacco companies funded research ostensibly demonstrating that smoking did not cause adverse health effects, all the while knowing that it did. Investigating who is paying for the expert's opinion is a particularly fruitful line of inquiry when such cynical manipulation of research is so common.

Some authority figures unknowingly promote the messages of self-interested institutions. In *Diet for a New America*, John Robbins marvels at the way that the dairy and meat industries have for decades promoted attitudes favorable to their products by contributing educational materials to schools. All of us who grew up in the 1960s and 1970s remember the advice from our teachers to drink milk and eat meat to promote good health, but those nice charts describing the basic "food groups" were part of a brilliant marketing campaign, in spite of how official they looked.

The flood of facts and falsehoods in this data-laden age puts greater responsibility on us to question "information" that appears to be from authority figures. Information found on the Internet is less subject to institutional validity

checks than are stories published by more established information sources like *The New York Times* or *The New Republic*, but any source can propagate nonsense. Proceed with caution, and always corroborate what you hear or read with multiple independent sources before taking action.

16

Loyalty to petrified opinion never yet broke a chain or freed a human soul. — **MARK TWAIN**

CONVENTIONAL WISDOM (SIC) FROM WELL KNOWN AUTHORITIES

"But what ... is it good for?" —Engineer at the Advanced Computing Systems Division of IBM, 1968, commenting on the microchip

"There is no reason anyone would want a computer in their home." —Ken Olson, president, chairman and founder of Digital Equipment Corp., 1977

"So we went to Atari and said, 'Hey, we've got this amazing thing, even built with some of your parts, and what do you think about funding us? Or we'll give it to you. We just want to do it. Pay our salary, we'll come work for you.' And they said, 'No.' So then we went to Hewlett-Packard, and they said, 'Hey, we don't need you. You haven't got through college yet.' " —Apple Computer Inc. founder Steve Jobs on attempts to get Atari and H-P interested in his and Steve Wozniak's personal computer

"This 'telephone' has too many shortcomings to be seriously considered as a means of communication. The device is inherently of no value to us." —Western Union internal memo, 1876

"The wireless music box has no imaginable commercial value. Who would pay for a message sent to nobody in particular?" —David Sarnoff's associates in response to his urgings for investment in the radio in the 1920s

"Who the hell wants to hear actors talk?" —H.M. Warner, Warner Brothers, 1927

"We don't like their sound, and guitar music is on the way out." —Decca Recording Co. rejecting the Beatles, 1962

"Heavier-than-air flying machines are impossible." —Lord Kelvin, president, Royal Society, 1895

"Professor Goddard does not know the relation between action and reaction and the need to have something better than a vacuum against which to react. He seems to lack the basic knowledge ladled out daily in high schools." —1921 *New York Times* editorial about Robert Goddard's revolutionary rocket work

"Stocks have reached what looks like a permanently high plateau." —Irving Fisher, Professor of Economics, Yale University, 1929

"Airplanes are interesting toys but of no military value." —Marechal Ferdinand Foch, Professor of Strategy, Ecole Superieure de Guerre, early 1900s

"Everything that can be invented has been invented." —Charles H. Duell, Commissioner, US Office of Patents, 1899

"Louis Pasteur's theory of germs is ridiculous fiction." —Pierre Pachet, Professor of Physiology at Toulouse, 1872

CHAPTER

15

HOW GUESSES BECOME FACTS

All data should be treated with skepticism. Here is one example of how official statistics get created, as recounted by Alan Meier, a long-time friend and colleague of mine at Lawrence Berkeley National Laboratory (LBNL):

> In 1987, Steve Greenberg and I wrote an article in an energy magazine about the rising amount of energy use that did not fit into the traditional categories.[35] As part of the article, we created three tables showing owner-ship of these small appliances (like fish tanks and power tools) and their esti-mated annual energy consumption. These values were based on very limited monitored data, back-of-the-envelope calculations, and hunches. The tables were assembled in one evening. (Many of the envelopes with calculations were then discarded.)
>
> In 1989, the U.S. Department of Energy's Energy Information Admin-istration (EIA) published its official *Household Energy Consumption and Expenditures*, including the results of their 1987 survey and additional analy-sis.[36] The EIA published the Meier and Greenberg data in a new table, "U.S. End-Use Consumption of Electricity for Selected Appliances." Whereas we published ranges in our estimates, the EIA just calculated and printed the averages from the high and low values. The word "estimated" appeared nowhere in the EIA table, so the reader was led to believe that these numbers were exact (curiously, the EIA is careful to give confidence bounds and other statistical parameters for its own survey data). To add insult to injury, the EIA misspelled my name in the citation.

29

Alan's example is more the rule than the exception. When little information is available about a particular topic, any moderately credible source gets cited by everyone concerned with the topic and becomes the new conventional wisdom. This happens frequently even though such estimates are often based on extremely crude assumptions.[37] I now sometimes joke that ALL estimates of energy use of appliances somehow originate with Alan.

Another colleague, Chris Calwell, was hired in 1996 to write a research report about the energy and safety problems with halogen torchieres, those inexpensive floor lamps that have become so popular in recent years.[38] Here is his story:

> In order to write the report, I needed to figure out how many halogen torchieres had been sold. The trouble was, nobody knew. The Census Bureau knew how many halogen bulbs had been imported but not how many were in fixtures. California utilities had counted how many were in a small sample of houses, but that was before the lamp had become popular. Market researchers had asked how many people bought halogen lamps of all types but didn't know how many were torchieres. The library was not much help, either, because only two articles had ever been published on this subject before I set out to write mine.
>
> I called the author of those two articles and got the names and numbers of her sources (the manufacturers). Then I called the manufacturers and asked how many they thought had been sold in total the previous year. I also asked them how average prices for the lamps had changed over time and about when they began selling the lamps in the United States. Taking all these different pieces of information under consideration, I created a table of sales of torchieres per year and the number of fixtures still in use at the present time. It was pure guesswork, informed by the information I could find, but guesswork nonetheless, with a fancy spreadsheet and graphs to back it up.
>
> The U.S. Consumer Product Safety Commission, a federal agency, was about to put out a press release on the problems with halogen torchieres at the same time I was finishing my report, but they also had no idea how many had been sold. They asked for my number (40 million), which my publisher reported to them as 35 to 40 million to be conservative. The CPSC thought that range seemed high, so it used 30 to 40 million in its press release to be even more conservative. During the next year, dozens of newspaper articles and TV programs cited the CPSC estimate of 30 to 40 million and attributed it to the federal agency, ignoring the original source and never bothering to examine the eight-line footnote in my original report documenting how the original estimate of 40 million was determined. The number had taken on a life of its own, proof of truth by institutional adoption and media repetition.

33 Your best defense against these sanctified guesses is to read the documentation for any data, including the footnotes. Track down cited sources and read them,

too. You should be able to figure out the methods used to create any data. If the documentation is not up to this task, you should regard the data with extreme suspicion. Don't ever use data unless you know how they were derived, you trust the cited sources, and you agree with the stated methods.

27

EXERCISE

Find an official statistic that sounds plausible and explore its origins. Do you still find it plausible after you've investigated?

The Government are very keen on amassing statistics. They collect them, add them, raise them to the nth power, take the cube root and prepare wonderful diagrams. But you must never forget that every one of these figures comes in the first instance from the village watchman, who just puts down what he damn pleases.

— **SIR JOSIAH STAMP,**
Inland Revenue Department of England,
1896-1919

CHAPTER

16

DON'T BELIEVE EVERYTHING YOU READ

Never act solely on information you read in the newspaper or hear over the Internet. News stories are often wrong either because the reporter misunderstood her sources or because the sources themselves were misleading or incorrect. Reporters sometimes report fiction as fact. Email claiming to contain valuable information can easily be a hoax.

One example is that of Stephen Glass, who wrote about 40 articles for *The New Republic* in the mid-1990s. Almost three-quarters of those stories were "partly or totally bogus," and the editor of that magazine was forced in mid-1998 to print a contrite mea culpa when Glass's deceptions were discovered.[39] The standard fact-checking process was not equipped to detect deliberate deception.

Some reporters even attribute quotations to those they have interviewed, when the person being interviewed never said any such thing. Erik Brynjolfsson, a professor at MIT's Sloan School of Management, had this experience in the spring of 1999. After I showed him a newspaper article in which he was cited, he said "I think I actually said about 25% of the quotes that were attributed to me. At least they didn't make me say anything too embarrassing." No one knows how frequently this happens, but it's common enough that you should not assume the veracity of quotations until you've verified them with the actual source.

Sometimes the sources themselves are at fault. Through the mid-1990s, the Software Publishers Association (SPA, now known as the Software & Information Industry Association, <http://www.siia.net>) periodically released estimates of the size of the software markets for Microsoft Windows and Apple Macintosh systems. In 1995 and 1996, the SPA proclaimed that Macintosh software sales had declined in both years relative to the prior year's estimates. The problem is that the SPA compared *preliminary* data in 1995 and 1996 to

final data in the previous year, but the final data in a given year contained substantial adjustments relative to the preliminary data (in the 1995 data, the adjustment was 45%!). The data series used by SPA was therefore not a consistent one, but the SPA's analyst felt confident enough to conclude that "the Mac platform is in decay." In fact, if final 1995 numbers are compared to final 1994 numbers, the real change in Mac software sales was about a 25% increase from 1994 to 1995, not a 14% decline as initially reported by the SPA. This result is not surprising, because Apple shipped record numbers of Macs in 1995.

The details of why the SPA made these appalling errors are not really that important for our purposes. The main lesson is that even respected institutions can distribute junk data to the world and pass it off as legitimate. If some Mac zealots had not delved into these numbers and compared the results year to year, unsuspecting software developers might have been led to the wrong conclusion. In fact, these widely reported estimates probably damaged Apple's business. How much damage they did no one can say for sure.

In the face of the public furor over its obvious mistakes, the SPA finally suspended its data program in September 1997. In hindsight, the SPA should have subjected its methodology to external peer review and been much more cautious **4** in its conclusions about Macintosh software sales.

The most famous book on ways to misuse numbers is Darrell Huff's *How to Lie with Statistics*. Huff documents dozens of examples where people, through incompetence or malicious intent, mislead readers with numbers, graphs, and other tools of the numerate. These tricks include biased samples, carefully chosen averages, graphs with chopped scales, and faulty logic in all its many manifestations.

You owe it to yourself to read Huff's book. It is justly revered both because of its accessible tone and its message: lying with statistics is commonplace, and it's wise to be aware of the most widely used tricks.[40]

What should you do to protect yourself and others from such shenanigans?

- First, *go back to the questions*. Read the questionnaire. Find out who **17** asked the survey questions and how they settled on the survey sample. Also determine when the questions were asked, because that can sometimes affect people's responses.
- *Be a detective* and determine the potential motives of the surveying **24** organization. Identify the funding sources for the research.

27

- *Dig into the numbers*, and compare results to other numbers you know to be true. Are results consistent with other sources? Can you find internal inconsistencies that cast doubt on the results? Answers to these latter two questions made it obvious that the SPA results were nonsense.

- Finally, don't email information to dozens of people unless you are confident it is true. I've received dozens of email warnings about computer viruses that will erase my hard drive or have some other terrible effect. I know most of these are bogus because they usually don't mention which kinds of computers they affect, and all genuine warnings are very clear about the software that is affected by a given virus. Viruses (with a few exceptions) afflict Macs, PCs, or UNIX machines, but one type of machine can't generally "catch" the virus from another type.

 Before forwarding ANY email about computer viruses, check out the latest information at the Symantec Anti-Virus Research Center, <http://www.symantec.com/avcenter/index.html>. For information on the most recent email hoaxes, see the Computer Incident Advisory Center's site at <http://HoaxBusters.ciac.org>.

Because of the ease with which information can be sent to dozens of people, news (or falsehoods) can spread quickly around the world. One example of this phenomenon is a collection of advice purported to be a commencement speech given by Kurt Vonnegut at MIT. The advice was witty, thoughtful, and altogether in keeping with Vonnegut's curmudgeonly but lovable take on the world. In fact, his own wife passed it around to her friends, thinking it was his work. It was widely distributed on email in the days following the MIT Commencement on 5 June 1997, and attributed to Vonnegut.

The essay included such gems as:

> Wear sunscreen. If I could offer you only one tip for the future, sunscreen would be it. The long-term benefits of sunscreen have been proved by scientists, whereas the rest of my advice has no basis more reliable than my own meandering experience. I will dispense this advice now.
>
> Do one thing every day that scares you.
>
> Sing.
>
> Don't be reckless with other people's hearts. Don't put up with people who are reckless with yours.
>
> Be kind to your knees. You'll miss them when they're gone.

Live in New York City once, but leave before it makes you hard. Live in Northern California once, but leave before it makes you soft.

Floss.

The only problem was that Kurt Vonnegut didn't speak at that MIT Commencement, nor did he write those delightful words of wisdom. The essay was a newspaper column by Mary Schmich, who writes for the Chicago Tribune. The column had somehow been stamped with Vonnegut's name and then sent merrily on its way to thousands of unsuspecting email readers.

This last example is a little reminder of how quickly and easily misinformation can be passed off as fact in the digital age. Keep it in mind as you venture forth into cyberspace.

EXERCISE

Find three popular articles on a topic you know a lot about. Make a list of the various misstatements you can identify based on your knowledge of the topic. Now go and read a few other articles on topics you know less about. Assume you find just as many mistakes. Worrisome? You bet!

The surest way to corrupt a youth is to instruct him to hold in higher regard those who think alike than those who think differently.

— **NIETZSCHE**

WHAT SCIENTISTS SAY, AND WHAT THEY MEAN:
A GUIDE FOR THE PERPLEXED

PHRASE	TRANSLATION
It has long been known	I haven't bothered to look up the reference
It is believed	I think
It is generally believed	A couple of other guys think so, too
It is not unreasonable to assume	If you believe this, you'll believe anything
Of great theoretical importance	I find it kind of interesting
Of great practical importance	I can get some mileage out of it
Typical results are shown	The best results are shown
3 samples were chosen for further study	The others didn't make sense, so we ignored them
The 4-hour sample was not studied	I dropped it on the floor
The 4-hour determination may not be significant	I dropped it on the floor, but scooped most of it up
The significance of these results is unclear	Look at the pretty artifact
It has not been possible to provide definitive answers	The experiment was negative, but at least I can publish the data somewhere
Correct within an order of magnitude	Wrong
It might be argued that	I have such a good answer for that objection that I shall now raise it
Much additional work will be required	This paper is not very good, but neither are all the others in this miserable field
These investigations proved highly rewarding	My grant is going to be renewed!
I thank X for assistance with the experiments and Y for useful discussions on the interpretation of the data	X did the experiment and Y explained it to me

SOURCE: Internet lore.

GO BACK TO THE QUESTIONS

It is common practice for companies to make decisions about marketing new products based on surveys of consumer behavior and preferences. In 1996, I spent about four hours filling out such a survey, and it got me to think about the limitations of such data.

The Survey of American Consumers is conducted by the American Institute of Consumer Studies.[41] An engaging young woman came to my door, explained what her job was, and started asking me questions. After about 15 minutes of questions she said that she had a booklet with more detailed questions, and that if I filled it out and returned it in the postage-paid envelope, I would be sent a check for $20.

The survey itself was 96 pages long, and it contained questions on every aspect of my purchasing behavior, from what kind of toothpaste I buy to how many miles I drive my car per year. The sheer endurance needed to fill out the survey was substantial, and I suspect that I was among the more thorough respondents. I dug through my cabinets to identify the brands of various foods and other household supplies. Sometimes I just didn't have the brand name but I knew I had used some form of the product during the specified time period, so I guessed (the instructions to the survey said to "estimate as best you can").

One thing that struck me was the number of times that the question did not allow me to explain my situation. For example, the survey asked about my newspaper reading habits ("Which sections of the daily newspaper do you read?") but did not allow me to explain that I read the *San Jose Mercury News* on line and did not at that time receive a hard copy newspaper.

Another question asked whether I use talcum powder. I do use it, but not to freshen up after a shower. Instead, I put it on ants that have invaded my kitchen; it drives them away, without using poison.[42] The makers of talcum powder might try to infer something from the results of this survey, but it tells them nothing about my non-conventional use of that product.

A worst case scenario...

Rhymes with Orange reprinted with special permission of King Features Syndicate.

Still another question asked where I do my food shopping but did not give me an option for the two local food stores at which I shop. Thus, it appears from my responses to the survey that I had only gone food shopping once in the previous 30 days even though I had eaten quite well during that period, thank you very much.

Given a sufficiently large number of respondents, a survey like this one will probably give results that reflect typical consumer purchases. Nevertheless, the anomalies I encountered should at least make you cautious about drawing firm conclusions from one survey without independent confirmation of the results.

These cautions are especially important when using "survey" data that can be skewed by so-called "self-selection bias." This problem commonly crops up when the institution conducting the survey has no control over who responds, which is typical for surveys conducted over the World Wide Web. Respondents with a particular point of view can easily "stuff the voting box," as *Byte Magazine* found when it tried to assess the prevalence of different computer operating systems.[43] The *Byte* survey results so clearly did not reflect the reality of computer ownership in the U.S. that the magazine discarded the results.

In spite of all these limitations, millions of business investment dollars are spent based on the results of such surveys. Before making important decisions based on survey data, you should examine the actual questionnaire used to collect the data and not simply accept the data as they are reported. The way a question is asked can affect results in substantial ways. This problem is accentuated if the data are broken down to show their locations within census tracts or zip codes because the small number of respondents inside each of those boundaries may lead to unrepresentative results.

The ways in which questions are asked, and to whom, can determine the out-

come of a poll. Crossen, in her book *Tainted Truth*, recounts a classic example from the early 1990s:

> Consider this question posed by Ross Perot. In a mail-in questionnaire published in *TV Guide*, the question was "Should the President have the Line Item Veto to eliminate waste?" 97% said yes. The same question was later asked of a sample that was scientifically [randomly] selected rather than self-selected, and 71% said yes. The question was rewritten in a more neutral way—"Should the President have the Line Item Veto or not?"–and asked of a scientifically selected sample. This time only 57% said yes.[44]

To truly understand the meaning of a survey, it is essential to go back to the original survey questions. *Never* base critical decisions on someone else's summary of survey results unless you have implicit trust in that person's judgment and understanding of the situation. It is also crucial to understand how the sample of respondents was selected (self-selected samples are the bane of a good analyst's existence). The sample must be truly *representative* of the population or else the results are suspect.

Real data always reflect the inherent inability of humans to track the changes happening all around us. That's why it's important to check new data with intuition, experience, and independent sources of confirmation before taking action. **3**

EXERCISE

If you encounter results of a poll that you find completely unbelievable, call up the firm that did the survey and ask to obtain the survey questionnaire. Try to determine whether misleading or unclear questions are the cause of the bizarre results.

Everybody believes in public opinion polls, from the man in the street to President Thomas E. Dewey. — **GOODMAN ACE, 1948**

18

READING TABLES AND GRAPHS

I confess! When reading a new report, I go straight for the tables and graphs. I do this for two reasons:

- I want to determine whether I find the author credible, and
- I like to understand the bottom-line results.

These two goals are quickly served by rummaging through tables and graphs for half an hour. If the tables and graphs are good enough, I read the paper to follow the author's reasoning more closely.

If the tables and graphs are poorly designed or confusing, I lose respect for the author. It is essential that tables and graphs summarizing analysis results be **33** clear, accurate, and well documented. If they aren't, including them is worse than useless because they hurt your argument and your credibility.

First check for internal consistency. I always begin at the bottom-line table and work backwards. I examine the column and row headings to be sure I understand what each one represents (I read the footnotes if I have questions). **27** I then assess whether the components of the total add up to the total. This procedure shows me whether the calculations are accurate, and it helps me become familiar with the various parts of the analysis.

It's a good idea to look over the numbers in the table and identify those that **27** are abnormally small or large. Typographical errors are quite common in tables (particularly in tables that summarize results from other, more detailed tables), and a quick scan can help you find them. For example, if you're reading a table summarizing hours worked per week by different team members on a project, an entry from one person that's ten times larger than the entries for others should catch your attention and prompt you to investigate further. The number itself may not be wrong, but checking it will increase your confidence in the numbers and help you understand how they were calculated.

Sometimes numbers don't add up exactly because of *rounding errors*, not because there is a mistake in the calculations. Say you've formatted a spreadsheet table so that there are no decimal places for the entries in the table. These entries might be 9.4 and 90.4, but in the table they are shown as 9 and 90, because the convention is to round numbers down to the next whole number when the decimal remainder is less than 0.5 (the remainder in both cases is 0.4 in this example). The sum is 99.8, which rounds to 100 and is shown in the spreadsheet as the total. The sum of 9 and 90 is 99, which makes the total of 100 look wrong even though there's a perfectly sensible explanation. If you're not aware of this potential pitfall, you could be misled by these apparent errors.

Read the footnotes carefully. They should convey the logic of the calculations and the sources of key input data. If you can't determine the methods used from the footnotes, you should be especially suspicious of the results and investigate further.

Check for ambiguous definitions and terminology. For example, there are at least five distinct definitions of the word "ton," and analysts often neglect to **33** specify which definition they are using. If it isn't crystal clear what the label means, you are likely to be led astray in interpreting the numbers. Another slightly different example involves the number of hours in a year. Many analysts assume it is 8,760 hours, but on average it is 8,766 hours because of leap years. This difference is essentially a definitional one, but it can lead to small inaccuracies in calculations.

The next step is to check consistency with independent external sources to make sure the data in the tables and graphs are roughly right. Compare growth over time to growth in other key data like population and Gross Domestic Product to get perspective on how fast something is growing relative **27** to these commonly used indicators.

Does the information in the tables or graphs contradict other information you know to be true? That table of hours worked may list Joe as someone who worked little on the project at hand, but if you know for a fact that Joe slaved over this project for many weeks because it was his idea, then you'll need to check the calculation. Similarly, if there's an entry of 175 hours for one week, it must be a typo or a miscalculation, because there are only 168 hours in a week.

Take ratios of results and determine whether the relationships they embody **27** make sense. If one component of the total is growing much faster than other

components over time, or if it is especially large compared to others, then investigate further. Look for large discrepancies and investigate when you uncover them.

Follow up when you encounter cognitive dissonance–any contradiction between your knowledge and the information in the table will lead to greater understanding, one way or the other. If there is a logical explanation for the contradiction, you've learned about the relationships between the information in the table and what you knew before. If the contradiction indicates a real problem, you've identified a flaw in the analysis. Root out the causes of cognitive dissonance, and you will enhance your knowledge without fail.

EXERCISE

Find several reports that you've either used or written over the past few years, and dissect the tables and graphs. Can you find all the information you need to use and to reference the results in the reports? How many unexplained discrepancies did you find? Have your tables and graphs improved over the years?

Although we often hear that data speak for themselves, their voices can be soft and sly.
— **FREDERICK MOSTELLER, STEPHEN E. FIENBERG, and ROBERT E. K. ROURKE**

DISTINGUISH FACTS FROM VALUES

Every human choice embodies certain values. When technical people advocate certain technology choices (e.g., whether or not to build more nuclear power plants) they often portray their advice as totally rational, completely objective, and value free. This portrayal *cannot* be correct—if an analyst makes a choice, that analyst has also made a value judgment. For example, Bernard L. Cohen[45] portrays his support for a revival of nuclear power as the result of scientific analysis when in fact it is a *public policy judgment* that is based both on his values and his scientific understanding.

Robert Reich's public policy students often forgot that values (and hence politics) are always intertwined with public policy choices:

> Some of my students used to regard public policy-making as a matter of finding the "right" answer to a public problem. Politics was a set of obstacles which had to be circumvented so the "right" answer could be implemented. Policy was clean—it could be done on a computer. Politics was dirty—unpredictable, passionate, sometimes mean spirited or corrupt. Policy was good; politics, a necessary evil.[46]

One purpose of analysis is to support public or private choices; in this context, it cannot be "clean" or value-free. It can, however, illuminate the *consequences* of choices so that the people and institutions making them can evaluate the alternative outcomes.

When Victor Fuchs et al. analyzed economists' positions on policy issues, they found that those positions were strongly correlated with each economist's values. Economists with a conservative political ideology viewed policy issues one way while those with a more liberal bent generally viewed them in a different way. Jonathan Marshall summarized the results nicely:

What the authors found was that economists' views on policy questions correspond more closely to their values than to the [facts].

To some extent, in other words, mere facts don't change their mind on issues. Keep that in mind the next time you see an ad, signed by 200 economists, supporting some candidate or piece of legislation. The ad may tell you less about their expertise than their ideology.[47]

I define facts as assertions about the physical world that can be verified through experiment, direct measurement, or observation. Events can also be facts although the interpretation of the meaning of those events is usually strongly colored by ideology and attitudes. Values, on the other hand, are explicitly subjective and are an expression of the ideas and feelings that are most important to us.

The relationship between facts and values can be mapped nicely on a two-by-two matrix (see **Figure 19.1**). If people agree about both the facts and the values related to a particular decision, reaching an acceptable solution becomes merely an issue of computation. A situation in which people concur about values but disagree about facts presents another easy choice: in this case, experimentation is the order of the day. If the facts are not in question, but people's values are in conflict, then they can negotiate to solve the problem. The most complicated and difficult choices are those where both facts and values are in dispute. In such situations, paralysis or chaos results. Most difficult public policy choices are problems of this type, and the best approach here is to experiment and bring the facts to a point where people no longer dispute them; people can then start to negotiate about values.

FIGURE 19.1. How society chooses[48]

	Agree on values	Disagree on values
Agree on facts	Computional decision	Negotiate
Disagree on facts	Experiment	Paralysis or chaos

In any of these discussions, it is a common rhetorical trick to link an ostensibly demonstrable fact with a value judgment. Such statements are of the form "Fact A is true, therefore we should take Action B." By linking facts and values in this manner, the speaker is explicitly or implicitly hiding the value judgment she has made.

Let the reader beware: Fact A may or may not be true, and Action B may or may not be a good idea whether Fact A is true or not. Too often these glib statements are allowed to pass without proper scrutiny. By being aware of this commonplace rhetorical technique (and by using your critical thinking skills), you can prepare yourself to respond.

The purpose of analysis is to help you illuminate the facts so that you can understand the choices of others and best use your values to make good choices of your own. Value choices are inescapable, which is why separating demonstrable facts from value judgments is so essential. Do so, and you are sure to be on a solid footing.

The significant problems we face cannot be solved by the same level of thinking that created them. — **ALBERT EINSTEIN**

THE UNCERTAINTY PRINCIPLE AND THE MASS MEDIA

One of Werner Heisenberg's major contributions to quantum theory was the so-called "uncertainty principle." This principle governs the relationship between the observer and the physical world on the subatomic level. To measure the velocity and position of a particle (an electron, for example) an observer must bounce something off that particle but by that action will change the electron's velocity and location. The more the observer tries to pinpoint the location of the electron, the more the measurement disturbs the velocity. Similarly, if the observer attempts to measure the velocity of the electron, its position becomes less and less certain. The level of certainty with which the observer can know the position and velocity of the electron is constrained by this physical law.

This principle can be loosely summarized for use in the everyday world as "the observer disturbs the system." Any time a measurement is made, that measurement changes the system being measured, usually in some miniscule way. Measuring the temperature of a liquid, for example, requires that a small amount of heat be extracted from the liquid and transferred to the thermometer. In physical terms, a good measurement is one that disturbs the system as little as possible. Things are a bit more complicated with social systems, especially when observations by influential members of the mass media can disrupt such systems in a significant way. The recent historical record is rife with both intentional and unintended disturbances of this kind.

The phenomenon of publicly available ranking systems is one example of how making new information available can lead to institutional change. Before 1984, when Milton Moskowitz and Robert Levering first wrote their book *The 100 Best Companies to Work for in America*, people had difficulty learning about family issues related to job choices because systematically compiled information was not yet available. When this work was published, the rankings became a force for change that had not previously existed.[49] The same effect

occurs whenever pollution emissions or campaign contributions are made public in an accessible way: embarrassment, education, or enlightened self-interest cause people and institutions to modify their behaviors (check out the Environmental Defense Fund's Scorecard web site at <http://www.scorecard.org> for a good example of how data and analysis tools can be designed to affect institutional behavior).

38

Another example of intentional disruption of observed systems is the set of opinion polls that influence the opinions they attempt to measure.[50] The pollsters ask questions of a carefully selected sample of individuals and then create a press release summarizing some of the key results. These "promotional surveys" are increasingly common. They can be conducted by a company for its own marketing, a non-profit institution for its own fundraising, or by a "neutral" third party like a trade magazine. For example, Kiwi Brands (a shoe polish company) polls people on a regular basis on such important topics as the relationship between ambitious people and shiny shoes and what the biggest grooming mistakes are: not surprisingly, badly shined shoes came out near the top.[51] Another company well known for this tactic makes sugarless gum ("Four out of five dentists surveyed prefer . . ."). In any case, read the fine print, and learn who was surveyed, how and when they were surveyed, and what questions they were asked. It is easy to create a survey that will give the desired result, and there's no way to know the truth without going back to the survey questions.

17

When the mass media report on companies wrestling with financial problems, they often cause unintended harm. Reporters repeat again and again the time-worn opinions of pundits, both well- and ill-informed. A whirlwind of bad publicity creates uncertainty and doubt among a company's customers, thereby influencing the very event on which the news people are reporting. The reporters create, in Cialdini's words, a kind of "social proof" that a company will fail, thereby making it more likely that it will.

An unintended result of polls can be that their results, taken out of context, often make headlines. People read the headlines and rethink their opinions based on them (if "thinking" is the right word to describe this process). The poll itself becomes news, and, in so doing, it influences the outcome of elections and changes public opinion.[52] The exact effect polls have on elections is uncertain, but clearly they can influence the results, as shown by the 1948 U.S. Presidential election. Crossen quoted Burns Roper (son of the founder of the Roper polling

organization) as saying that the 1948 polls "lulled the Republican supporters into a false sense of confidence and scared the bejesus out of organized labor." The pollsters also believed that people had decided on their presidential choice weeks before the election, so they stopped regular polling. In addition, telephone polls only reached people with telephones, who were generally wealthier than the population at large and were therefore more likely to vote Republican. One famous result was the erroneous *Chicago Tribune* headline "Dewey defeats Truman!"

Mass media amplify the effect of such observations, and this tendency will only become more prevalent as the cost of transmitting information continues to plummet. Corroborate mass media reports with independent information before using them or passing them to others. If you repeat such reports without analysis, you enlighten no one and may cause harm to people and institutions that you do not intend. Finally, do your utmost to ensure that your own research results are reported correctly so that you yourself do not fall prey to this pitfall.

Intro

Public opinion is a compound of folly, weakness, prejudice, wrong feeling, right feeling, obstinacy, and newspaper paragraphs.

— **ROBERT PEEL**

CREATE YOUR ANALYSIS

Now that you've gleaned what you can from the analysis of others, it's time to roll up your sleeves and get to work. Each of these short chapters offers key insights to help you get started. The order in which you read them is not so important, so first choose the ones that interest you most, but do read them all before you're through.

It is a mark of educated people, and a proof of their culture, that in addressing any problem they seek only so much precision as its nature demands or its solution requires.

— **ARISTOTLE, in *NICHOMACHEAN ETHICS***

CHAPTER

21

REFLECT

Any successful problem solver develops an array of techniques to spark her creative process. Bryan Mattimore describes dozens of such techniques in his book *99% Inspiration*, every single one of which involves reflection on the nature of the problem and the possible solutions.

This reflection can be conscious, unconscious, intellectual, intuitive, planned, unplanned, or all of the above. The key is to give yourself time to reflect on the problem and let a solution bubble to the top of your consciousness.

Reading books like Mattimore's can be helpful, but each person must develop techniques that mesh best with her own problem-solving style. When I am faced with a particularly thorny issue, I go for a walk or a run, which stirs my creative juices and removes me from other distractions. You should do what works for you: many people listen to music, make pottery, take a bath, or clean house to attain the focus they need for true reflection.

I begin my reflections by listing what I know about the problem and what a satisfactory solution would entail (without knowing the exact solution, you can still define the kind of information that would solve the problem at hand). I then try alternative approaches in my mind, analyzing each one in terms of likely effort expended and probability of success, considering the available data. I let my mind wander as I walk, which often leads to important insights. By the end of my walk (assuming I haven't been hit by a car during my reveries) I have at least one possible way to attack the problem.

It is a rare person who can think creatively in the face of constant interruptions. I can't do it, and I suspect most others can't either. Put aside some time every day to reflect a bit. Turn off the phone, close the door, and put up a "Do not disturb" sign. Schedule this reflection period during your particularly productive times of day.

Brenda Ueland, in her book *If You Want to Write*, asserts that creativity only bursts forth when we are idle and thinking on our own, uninterrupted. The con-

stant busyness we often experience in modern life just gets in the way of our own creativity. That's why she advocates at least half an hour per day of this sort of idleness.

Joe Hyams, in *Zen in the Martial Arts*, reports how one martial artist (Bronislaw Kaper) takes this idea to its logical extreme and consciously schedules entire days to do nothing. By taking time to reflect, you too can use such pauses to your advantage.

Music is the space between the notes. — **CLAUDE DEBUSSY**

22

GET UNSTUCK

Sometimes, try as I might, I just can't push past a seemingly insurmountable obstacle. The problem itself may be especially difficult, or it may pose personal challenges. Any successful problem solver develops tricks for "getting unstuck" in the face of such obstacles.

One of my most successful tricks is to break my normal routine. Some of my best insights have come on airplanes where I have time to reflect and have few of my normal distractions. Go for a walk or a run, ride your bike, or go on a trip. Call a friend you haven't seen in a while. Do something out of the ordinary and forget about the problem for a few hours. Brilliant insights often occur when you're relaxed and not thinking about the problem at all.

You can also try working on a related problem or one that's entirely different. Either way, you'll be able to start fresh with the more vexing problem after you've attacked something a bit more tractable.

An artist I met once described how, once he had an idea for a painting, he would start on the canvas immediately. Once he got "inside the picture," the difficulties would often just melt away. The same technique works for other kinds of problem solving. If a problem remains daunting, ask if there's some small question that you can answer (jump "into the picture"). Answering that question often results in a bunch of obvious follow-up questions that can also be answered, and addressing these usually puts you on the road to a solution.

A time-tested technique for breaking logjams is to describe the problem to someone else. One friend recalled for me how her mother helped her solve math problems in junior high school. "Mom couldn't remember the details about specific math techniques, but suggested that I explain the problem to her. By the time I had given her the details, I had virtually always had a new insight for a solution. Explaining the problem forced me to think clearly."

The converse of this approach is also true, which is that clear and precise explanations of results indicate clear and precise thinking on the part of an ana-

lyst. Those who can't explain their own analysis in an understandable way are not clear thinkers, and you should steer clear of their work.

There is a series of questions that flow through my mind whenever I'm faced with a problem unlike any I've seen before. These questions, largely based on G. Polya's book (*How to Solve It*), are as follows:

- "What is the unknown" (i.e., the number I am trying to calculate)?
- What data are given, and which can be estimated using knowledge I already have? Are the given data correct based on other knowledge I have? Are there any additional data I can bring to bear on the problem?
- Can I restate the problem in different words?
- Can I draw a picture or graph framing the problem?
- Is there another similar problem I have already solved that could give me the key to a solution? Can I solve a related problem?
- Can I make some simple assumptions that might help me solve the problem more easily? Can I solve just one part of the problem? Can I solve a simplified version of the problem, using round numbers that are roughly right?
- Are there any answers that I know are wrong? (Sometimes you can get closer to an answer by eliminating those that cannot be correct.)

24

The study of problem solving, which Polya calls "heuristic," is still a nascent art. Each obstacle is an opportunity to create and refine your own heuristic process. Examine your problem solving style, and create techniques for getting unstuck that work for you.

EXERCISE

Ask people whose judgment you respect how they get unstuck; then, try their techniques. Were you surprised at any of their answers?

A mistake is an event that has not yet been fully turned to your advantage. — **EDWIN LAND**

CHAPTER

23

INQUIRE

Quick—your boss just assigned you a task that is outside your area of expertise. What do you do?

I am often faced with this challenge, as are most analysts in business, academia, and government. Because the problem is new, it's easy to have a "Beginner's Mind." My first move is always to pick up the phone. I have friends [1] with a wide range of interests, and usually there is at least one with direct knowledge of a particular topic. By asking these people focused questions, I can quickly determine what is possible within the time allotted. I often develop these questions as part of my effort to "Get Unstuck." [22]

Don't be afraid to approach experts with questions. They are often delighted that someone is taking an interest in their work and will usually bend over backwards to explain why their results are useful and important. The reason many people become experts is because of a passionate interest or concern. They *care* about their topic and want other people to care too. So go ahead and call them up. Your interest will make them happy, and you might just learn something along the way!

"Thank God! A panel of experts!"

9 If you don't know anyone who knows about the topic, try the Internet. Use one of the many search engines available for this task (try several, because they are all virtually guaranteed to give different results). Some search engines (e.g., <http://www.ask.com>) allow you to ask questions in plain English and receive focused results that number ten or twenty links (instead of the thousands you normally get from such searches). To hone your Internet research skills, consult some of the recent work that teaches these skills (e.g., <http://dir.yahoo.com/Computers_and_Internet/Internet/World_Wide_Web/Searching_the_Web>).

As discussed earlier, an important tip in Internet research is to consult the Frequently Asked Question sites (FAQs) that exist on the Internet for a truly amazing number of topics <http://www.faqs.org>. FAQs are compiled by experts and are often a superb place to begin your research. Use one of the many excellent Internet search engines to find FAQs that are relevant to your inquiry.

The web site <http://www.Amazon.com> is a surprisingly good place to do research. The main purpose of this site is to sell books, but its search routines are easy to use, and the site presents for each book a list of other books commonly purchased by people who bought the featured book. It also allows you to see other books by the same author with one click of the mouse. There are millions of books in the Amazon.com database, which is more than enough to cover most topics thoroughly. Even out-of-print books are included. I spent an instructive couple of hours exploring the site early one morning, and I came away impressed.

Another place to begin is a major reference work. For example, you can find **9** something related to virtually any query in *The Statistical Abstract of the U.S.* or in an encyclopedia. The general information contained in these works can often be enough to get you started untangling even the thorniest research problem. For many social and physical science topics, check out the many different **9** *Annual Reviews* that summarize the latest work in these fields.

If you are still stymied, talk to the reference librarian at the local library. Librarians have an uncanny knack for tackling just this sort of inquiry, and they are highly skilled professionals. They also love the challenge of finding new and interesting information. At a minimum, the librarian should be able to point you to the standard reference work in any area of knowledge as well as to the relevant trade journals (which can be treasure troves of information about real-world markets and technologies).

Librarians play a crucial role in research. The importance of these professionals came to the fore in a reporter's tale of searching for information in the spanking new Main Library in the city of San Francisco.[53] The reporter tried searching for books on a probably mythical Christian king who was the impetus for several important missions of exploration during medieval times. Although his topic may seem obscure, the reporter's experience will ring familiar to most. He tried first to use the library's electronic card catalog and came up with two listings on his subject, one of which seemed irrelevant. On the advice of the reference librarian, he then logged onto the library's on-line database and found seven new listings that didn't show up in the first search. Then he found another new listing by changing the order of his search terms from "Prester John" to "John, Prester." Finally, he consulted the old card catalog that the computer terminals had largely displaced and found an additional reference (a book that was so old that it hadn't been entered into the computerized database and was kept not in the main collection but in a storage annex).

Without the help of a librarian, this reporter might have missed some key sources. Only a librarian could tell him the tricks that revealed the sources he sought in an expeditious manner. Trial and error can yield results, but your search will be far quicker if you consult a librarian first.

Working outside your area of expertise is difficult, but can be rewarding. As we know from the history of science, many of the greatest discoveries were made by scientists who were formally trained in another discipline. Developing the skills to explore new fields will serve you well in any endeavor you undertake.

Minds think with ideas, not information. No amount of data, bandwidth, or processing power can substitute for inspired thought.

— **CLIFFORD STOLL**

24

BE A DETECTIVE

One of the defining features of the literary character Sherlock Holmes is his scientific approach to solving mysteries. He has an immense store of knowledge of past criminal activity upon which he can rely. He systematically collects information from the scene of the crime and notices the most seemingly trivial details that might bear upon his case. He does not prejudge the case in advance of the facts. Finally, he is skilled in weaving alternative theories that might fit the facts.

Holmes' approach is not unlike that required to solve problems in the modern world. Any problem solver worth her salt knows as much as possible about data and previous work on the problem at hand (if you're new to it, then go the library and read). Attention to detail is also essential, as is not allowing your initial snap judgments to color your interpretation of parts of the puzzle. And weaving alternative theories to fit the facts is perhaps the most important skill of a seasoned problem solver.

Spreadsheets[54] are my favorite problem-solving tool because they allow me to create many different scenario calculations easily and quickly. Business people use them to create financial projections for their business plans. Scientists use them to analyze data from their experiments. Policy analysts use them to project the effects of proposed government policies on people's lives and on the economy as a whole. In each case, spreadsheets enable both quick scoping calculations and extremely detailed analysis.

Sometimes in creating a spreadsheet you discover an intermediate or final result that is either internally inconsistent with another such result from the same calculations or that cannot be right based on first principles. An example of the first kind of inconsistency would be to have a business plan that showed operating losses for the first five years of the forecast but that showed no expenditures of startup capital after five years (the company would need to use the startup capital to pay for the operating losses until the company became profitable). An example of the latter inconsistency would be to project that a small

startup company would gain 99% market share within a few years and attract no competitors (any time an important business niche appears, competitors almost always pop up to contest it).

When you discover such inconsistencies in your spreadsheets, Holmes' technique of systematically investigating alternative theories comes into play. Begin by making a list of possible reasons to explain why the calculation is in error; then, trace the flow of data through the calculations, starting from the point at which you noticed the inconsistency. Look at each and every input to a particular calculation to see whether you've made any obvious errors in entering formulas or data; trace the calculations all the way back to the beginning. If, as is commonly the case with complicated calculations, you still can't figure out what's wrong, recreate your calculation in a new spreadsheet, usually using a simplified version of the calculation and plugging in data that are roughly right (and that are nice round numbers like 100 or 1,000). This approach allows you to begin with a clean slate. Often when you are immersed in calculations you can forget small details. Starting fresh in this way helps you catch those errors.

One quick check you can do on your spreadsheet calculations is to vary a key input number, and see if the results change as you expect. If you increase that number by 50%, do you expect the result to increase or decrease by a comparable amount, or by half that amount? If you make that number close to zero or extremely large, do the results behave as anticipated? If your results vary in an unexpected way or in unusual proportions, your calculation is wrong or your conceptualization of the problem is flawed in some way. Either way, you've learned something.

If two complex calculations should yield the same result but do not, break down each calculation into its component parts, and compare each part of the first calculation to the corresponding one in the second. This technique will make it obvious where in the calculations any differences lay. For example, assume that a formula in a spreadsheet has three terms that add to a total (Total = A + B + C) and that an independent calculation has three terms that should also add to that total (Total = X + Y + Z). Assume also that the value of each term in the first calculation should correspond to the value of each term in the second calculation (A should equal X, for example). If the result of the two calculations is different, then compare A to X, B to Y, C to Z, and see if the differences in the overall total can be explained by differences between any of the individual components. You can usually narrow down the discrepancy to one

or more of the terms in the equation and eliminate some possibilities. Although tedious, this technique can help you identify the potential explanations for such discrepancies. It is especially useful when each of the terms is, by itself, a complex calculation.

When working with spreadsheets, many people copy and paste data from one sheet into another. This practice is a pernicious one because it makes updating your calculations later a difficult task and makes it nearly impossible to determine how the calculations flow, once they are no longer fresh in your mind. Obey this golden rule: *Never paste values from one spreadsheet into another*. Link spreadsheets together directly so that when you change your calculations in the first spreadsheet, the other one is automatically updated. Following this rule helps you track down errors in the calculation much more quickly. It also saves you time if you have to change input assumptions later, and it creates a special kind of documentation of your calculations.

If your results look too good to be true, they probably are. Be suspicious of your own calculations when you first create them. *Dig into the numbers* to determine whether you can trust them.

Another common technique used by detectives is to consider the motives of the person or institution putting forth a particular viewpoint. I suspect, for example, that a lobbyist for the tobacco industry might be prone to overlook evidence that smoking is harmful. Seek out sources whose interests coincide with the truth, and discount somewhat the opinions of those with a vested interest. However, remember that having an interest or an opinion is not evidence of being biased any more than *not* having an opinion is evidence of being unbiased: some people (a rare few) are honest even to their own detriment.

When you have eliminated the impossible, whatever remains, however improbable, must be the truth. — **SHERLOCK HOLMES**

CREATE CONSISTENT COMPARISONS

In the real world, so many things change at once that it is often difficult to decipher cause and effect. One of the most compelling techniques for making a point in the face of this complexity is to find or create a consistent comparison that eliminates the complicating factors and illustrates your point in an unambiguous way. A pure consistent comparison is one where the only feature differing between two examples is the one that illustrates your point. Even when there are some minor differences, you can often correct for these and still create a powerful comparison.

Isolating the root cause by creating an experiment is one of the most fundamental techniques of modern science. When medical researchers create "control groups" to compare to people being given an experimental drug, they are creating a consistent comparison or as near to one as is humanly possible. Although you can't always create such an experiment, you can often uncover one if you look carefully. I discovered such a case in the early 1990s, which led me to write one of my favorite papers. Since the 1970s, U.S. automakers had insisted that they were unable to increase the fuel efficiency of passenger cars in a cost-effective way. In 1992, Honda introduced a new Civic hatchback that was the same size and had the same engine power as the model it replaced, but it was 56% more fuel efficient than the previous year's model. I read news articles about this new car and badgered various friends who work on transportation issues to investigate this case further. After several months of fruitless nagging, I decided to explore it myself, with the assistance of two colleagues from the Union of Concerned Scientists.

The two cars differed slightly in features but were essentially identical except for their fuel economy. The case was therefore a good one to use to estimate the actual retail costs of improving fuel economy in the real world. After correcting for the costs of the different features (based on dealer cost and list prices), the more efficient Civic (the 1992 VX) cost about $700 more than the

1991 Civic DX, and saved about 110 gallons of gasoline per year for typical drivers. The added cost would pay for itself in about five years at then prevailing fuel prices, making it a cost-effective investment.

The example was an exceptionally powerful one because the size and engine power of the two vehicles were identical, and the feature list was nearly so. As a check on the calculation of the retail cost, we also investigated the projected retail cost of the technologies used to improve the VX's fuel economy, based on *independent engineering estimates* and found that these estimates were quite close to our calculation of the actual retail cost difference between the 1991 DX and the 1992 VX. This independent check gave us confidence that we were on the right track.

The 1993 Civic VX gets 55 miles per gallon on the freeway yet holds four passengers comfortably. It's not a luxury vehicle by any means, but it's a testament to what automakers can do if they put their minds to it. The paper documenting this case subsequently won a prize from the National Research Council, which is an indication of the compelling nature of a well-chosen consistent comparison.

The cure for boredom is curiosity. There is no cure for curiosity.
— **DOROTHY PARKER**

TELL A GOOD STORY

Life is about making choices in the face of uncertainty. One oft-neglected [3] approach to such decision making is the art that Peter Schwartz calls "scenario building." The purpose of scenario analysis, as explained by Schwartz in his now classic book *The Art of the Long View,* is to explore several possible futures in a systematic way. No matter how things turn out, the analyst will have "on the shelf" a scenario (or story) that resembles that future and will have thought through in advance the likely consequences of that scenario on the decisions at hand. Such a story "resonates in some ways with what [people] already know and leads them from that resonance to reperceive the world."

Schwartz describes eight steps for successful scenario development; I summarize them here:

- *"Identify [the] focal issue or decision"*: Scenarios should be developed in advance of some major strategic decision so that when the time to decide arrives, the options are laid out in detail and the implications of each decision path are made manifest. Such a decision might be "Should our company expand into new markets next year, or should we focus on expanding our market share in our current markets?"

- Determine the *"key forces in the local environment"*: The next step is to identify the key factors determining whether the focal decision succeeds or fails. These factors might include such things as customer acceptance of a new technology or competitor response to the new strategy.

- Identify the *"driving forces"*: This step involves identifying the forces in the larger society that affect the key local forces in step 2. These forces could include demographics, inflation, commodity prices, technological change, public policies, or political instability. Some of them are beyond the control of any individual or institution ("predetermined elements"), and others are subject to human influence.

- *"Rank by importance and uncertainty"*: Of the trends and driving forces identified in steps 2 and 3, a small number will be both highly important and highly uncertain (see also Morgan and Henrion's book *Uncertainty*). These are the issues upon which differences in the scenarios will be based. Predetermined elements don't differ among scenarios.

- *"Select a scenario logic"*: The most important and uncertain factors should then be used and varied to define the set of scenarios to be considered. In essence, these factors are the dimensions of uncertainty that the scenarios will explore. Schwartz gives the example here of an automobile company that decides that fuel prices and protectionism are the two main drivers of their company's future. The scenarios they consider would be: "(1) high fuel prices in a protectionist environment—where domestic suppliers of small cars will have an advantage; (2) high fuel prices in a global economy—where fuel-efficient imports may capture the low end of the market; (3) low fuel prices in a protectionist environment—where American gas guzzlers will have a good market at home but not abroad; (4) low fuel prices in a global economy—where there will be intense global competition for fuel-efficient models but larger cars may enjoy strong foreign markets."

- *"Flesh out the scenarios"*: Each of the key driving forces and factors from steps 2 and 3 should be assessed for each of the main scenarios. Usually, these factors can be assigned relative rankings (i.e., "Inflation should be higher than average in scenario 1 and lower than average in scenario 2"), which can then be translated into actual numbers. The exact numbers are not so important as long as they reflect in a credible way people's expectations about how the future might unfold in a particular scenario. The scenario must then be woven into a plausible story describing a course of events that would result in the scenario's final outcome. Each scenario must be named in a way that vividly reminds people of its essence.

- *Determine the implications*: Revisit the decision identified in step 1, and assess how that decision would fare in each of the different scenarios. If the decision (e.g., expanding into new markets) would lead to success (as defined in step 2) for all or most of the scenarios considered, that decision is probably a wise one. If it is likely to be successful in only one scenario, the decision is "a high-risk gamble . . . especially if the company has little control over the likelihood of the required scenario coming to pass."

- *"Select the leading indicators and signposts"*: Once the scenarios have

been developed, it is crucial to identify a few key indicators by which decision makers can monitor the evolution of events. These indicators are used to determine which of the different scenarios best matches up with actual events and can be used to trigger an institutional response (e.g., "if inflation gets up to 7% and the stock market goes down 15%, then do X"). If you can't measure it, you can't manage it, so defining appropriate indicators is crucial to success.

Schwartz believes that scenarios are successful when "they are both plausible and surprising; when they have the power to break old stereotypes; and when the makers assume ownership of them and put them to work." The institutional process of creating scenarios (what Schwartz calls "holding a strategic conversation") can also reveal new opportunities for an organization and promote team building among its employees.

Another name for scenario analysis is the commonly used scientific technique of creating a "thought experiment." All scientists use this technique, in which they imagine certain real-world complexities to be absent and then think through the consequences for the situation they are investigating. For example, Galileo imagined a world in which the complicating effects of friction did not exist and, by doing so, was able to make great strides in what would come to be called the field of kinetics (the study of moving objects). Such thought experiments are as valuable in business as in science. Instead of limiting your imagination by current circumstances and constraints, imagine a world in which some key constraints don't exist, and ask yourself how you or your institution might respond.

Most of Schwartz's examples of scenario analysis have only a small quantitative component, but many other futurists are obsessed with numbers and computer models. Most explorations of the future with which I am directly familiar err by focusing too much on the mechanics of forecasting and quantitative analysis (e.g. on particular modeling tools) and far too little on careful scenario development based on the procedures described above. Quantitative analysis can lend coherence and credence to scenario exercises by elaborating on consequences of future events, but modeling tools should support that process rather than drive it, as is so often the case.

In my years of involvement with development of national energy policy, I have been most struck by how few resources are devoted to sensible scenario

30

development and associated data and how many are devoted to the development of different modeling tools to assess such policies. Computer tools are sexy and appealing (at least to the funding agencies). Data and scenario analysis, upon which the results generally hinge, are virtually always given short shrift.

Tens of millions of dollars are spent every year on models whose capabilities are redundant with others, usually because a particular agency with money wants its own "in-house" modeling capability and is unwilling, because of institutional rivalry or personal biases, to adopt a pre-existing framework. Policy makers fail to realize that models are ALL unable to predict the future in an accurate way and that small improvements in modeling methodology are made irrelevant by inadequate scenario development.

The best way to predict the future is to invent it. — **ALAN KAY**

DIG INTO THE NUMBERS

Anyone who has delved into data from the real world knows that they're messy. Survey takers can write down the wrong response. People entering data from a survey form into a computer can type in the wrong numbers. Computers sometimes garble data because of software bugs (especially when converting one file format to another). Data formats become obsolete as software changes.

Electronic data recorders break or go out of adjustment. Analysts mislabel units and make calculational errors.

"Cleaning the data" to correct for such problems is virtually always necessary, but this step is often ignored. Always ask about the process used to clean the data: if the response is just a blank stare or worse, you know you're in trouble. Many companies (like AT&T and Sega of America) now assign people to check for data quality in the face of all the problems associated with real data.

Some of the more famous examples of how small mistakes in data processing can have disastrous results are found in the space program. The Mariner I spacecraft went off course soon after launch and had to be destroyed. The cause of the malfunction was a missing hyphen in a single line of Fortran code. Arthur C. Clarke quipped later that this $80 million launch was ruined by "the most expensive hyphen in history."[55] More recently, NASA's Mars Climate Orbiter was lost in space because engineers on the project forgot to convert English units to metric units in a key data file, a mistake that cost scientists years of work and taxpayers $125 million.[56]

Bad data can have a real cost for a business. If a 500,000-piece mailing uses an address list with an error rate of 20%, the company wastes $300,000 by sending mailings to incorrect addresses. It loses even more money because among those 100,000 missed prospects are about 1,500 people who would have become customers and purchased thousands of dollars worth of the company's products over their lifetime. The losses from these missed customers can be many times greater than the immediate direct losses.[57]

It is crucial to pore over raw survey data to check for anomalies before doing extensive analyses. For example, typographical errors can lead numbers to be ten, or a hundred, or a million times bigger than they should be. Looking over the raw data can help you identify such problems before you waste time doing statistical analysis using incorrect numbers.

Bad data ruin your credibility and call your work into question. Even if there's only one small mistake, it makes your readers or listeners wonder how many other mistakes have crept into your analysis. It's difficult to restore your credibility after some obvious mistake is revealed, so avoid this problem in the first place. Dig into your numbers and root out these problems *before* you finalize your paper or talk.

SOME SPECIFIC ADVICE

If you have data to be analyzed, type them into the spreadsheet yourself, assuming there aren't many hundreds of them. There's no substitute for this effort, even if you are a highly paid executive, because it will help you identify inconsistencies and problems with the data and give you ideas for how to interpret them. This technique also gives you a feeling for the data that can't be replicated in any other way. You will almost surely see patterns and gain unexpected insights from this effort.

Check that the main totals are the sum of the subtotals. Most documents are rife with typographical errors and incorrect calculations. You should therefore not rely blindly on any data source's summations but calculate them from the base data. You can check your typing accuracy by comparing the sums to those in the source of data. If they match exactly, it is unlikely that your typing is in error. Even if you don't check these sums, you can bet that some of your readers or listeners will. Do it yourself, and avoid that potential embarrassment.

Check that the input data are current. Don't forget that official statistics get revised all the time. Make sure you know the vintage of the input data used in the analysis. For example, don't compare analysis results generated using one year's Census data with those based on another year's data (unless your sole purpose is to analyze trends over time).

Check relationships between numbers that should be related in a predictable way. Such comparisons can teach valuable lessons. For example, when examining data on carbon emissions of different countries, a newcomer to the field of greenhouse gas emissions analysis might expect that the amount of carbon emitted per person would not differ much among industrialized countries. In examining such data, however, we find large differences in carbon emitted per person, from less than 1 metric ton/person/yr in Portugal to more than 6 tons/person/yr in Luxembourg. Determining why such differences exist is the logical next step, which will inexorably lead to further analysis and understanding.

Check that you can trace someone else's calculation in a logical way. If you can't, you can at least begin to list the questions you need to answer to start

tracing the calculation. Ultimately, if you can't reproduce the calculation, the author has broken a fundamental rule of good data presentation, and his analysis is suspect.

Compare the numbers to something else with which you are familiar, as a "first-order" sanity check. These comparisons can show you whether or not you are on the right track. Presenting such comparisons in reports and talks can also increase your credibility with your readers or listeners because it shows that your results "pass the laugh test."

Normalize numbers to make comparisons easier. For example, the true size of total U.S. Gross National Product (GNP) in trillion dollars per year is difficult to grasp for most people but if normalized to dollars per person per year will be a bit more understandable. Common bases for such normalizations are population (per person/per capita), economic activity (per dollar of GNP), or physical units of production (per kilowatt hour or per kilogram of steel produced).

If you have data that extend over time ("time series data"), normalize them to a base year to enhance comparisons. By expressing such data as an index (e.g., 1940 = 1.0) you can compare trends to those of other data that might be related. For example, if you plot U.S. raw steel production (**Figure 27.1**), population (**Figure 27.2**), and GNP (**Figure 27.3**) in separate graphs, it is difficult to gain perspective on how fast steel production is changing over time relative to these other two important determinants of economic and social activity. However, if you plot steel production over time as an index with 1940 = 1.0 (see **Figure 27.4**), you can plot population and GNP on the same graph. Such a graph will show instantly whether growth rates in these data differ.

In this example, real GNP grew by a factor of more than five from 1940 to 1990. U.S. steel production roughly doubled by 1970 and then declined by 1990 to roughly 50% above 1940 levels. Population in 1990 just about doubled from 1940 levels.

The trends for U.S. steel production in Figure 27.4 illustrate dramatically the changing fortunes of steel in a postindustrial economy. Just after World War II, steel use per capita increased as automobile ownership expanded and use of steel for bridges and other forms of construction also increased. After 1970, the

FIGURE 27.1. **U.S. production of raw steel 1940-1990 (million short tons)**[58]

FIGURE 27.2. **U.S. population 1940-1990 (million people)**[59]

FIGURE 27.3. **U.S. gross national product 1940-1990 (billion 1982 dollars)**[60]

FIGURE 27.4. **U.S. raw steel production, population, and gross national product 1940-1990, expressed relative to 1940 levels (1940 = 1.0)**

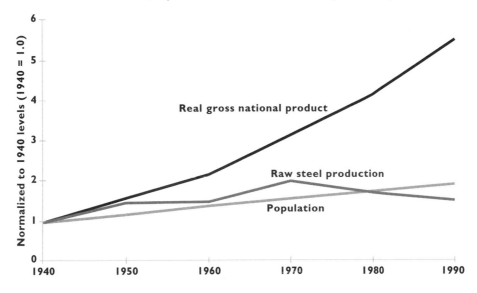

Real gross national product

Raw steel production

Population

steel industry began to face serious competition from foreign steel makers as well as from alternative materials like aluminum alloys and composites, so U.S. production declined even as population increased and real GNP went through the roof.

Break problems into component parts. Explore analysis results by examining the factors that lead to those results. For example, suppose someone tells you that the market capitalization of Microsoft was $430 billion and that of General Electric (GE) was $370 billion (numbers are current as of April 1999). What steps should you take to understand what these numbers mean?

Market capitalization is the product of the number of shares outstanding and the stock price per share, as shown in the equation below:

$$\text{Market Capitalization (\$)} = \text{\# Shares} \times \frac{\text{Stock Price}}{\text{Share}}$$

The stock price per share can be further broken down into the product of the earnings per share and the price-to-earnings ratio, yielding the following simple model:

$$\text{Market Capitalization (\$)} = \text{\# Shares} \times \frac{\text{Earnings}}{\text{Share}} \times \frac{\text{Price}}{\text{Earnings}}$$

The product of the number of shares and the earnings per share gives the total annual earnings (profits) for each company. Substituting in the previous equation, we get:

$$\text{Market Capitalization (\$)} = \text{Total Earnings} \times \frac{\text{Price}}{\text{Earnings}}$$

If we divide both sides of this equation by annual revenues, we get

$$\frac{\text{Market Capitalization}}{\text{Annual Revenues}} = \frac{\text{Total Earnings}}{\text{Annual Revenues}} \times \frac{\text{Price}}{\text{Earnings}}$$

All of these equations represent variations on the same model. Different forms of the model will be useful at different times.

For most companies, the basic data for the model are readily available on the

web, so that's the best place to start (you could also go to the library). **Table 27.1** summarizes the key financial parameters for calculating market capitalization for the two companies.

The first thing to notice about the financial statistics for these two companies is a huge disparity: Microsoft's market capitalization in April 1999 was about 20% larger than that of GE in spite of revenues that were only one-seventh as large. All other things being equal, we might expect that companies with similar market valuations would also have similar revenue streams. All other things are *not* equal, however, and finding out why will help illustrate this important analytical technique.

If Microsoft's market capitalization per dollar of revenues was the same as that of GE, we'd expect that its market capitalization would be only one-eighth as large as it is.[61] We need to explain this eightfold discrepancy, and to do so, we examine the components of the last equation above. The first component is earnings per dollar of revenues, and the second component is the price-to-earnings ratio.

As Table 27.1 shows, Microsoft's earnings are about four times greater than GE's per dollar of revenues, which accounts for about half of our eightfold difference. The price-to-earnings ratio also differs between the two companies. Apparently, the stock market values one dollar of Microsoft's earnings twice as much as one dollar of GE's earnings, which accounts for the remaining factor of two.

By breaking these numbers down into their component parts, we've been able to isolate the two key reasons for the eightfold discrepancy identified above. The next step is to explain why these two reasons exist.

Microsoft's greater profitability per dollar of revenue reflects the difference between industrial manufacturing and software development. The latter has low marginal costs of reproducing the product, and high gross margins. Microsoft's higher price-to-earnings ratio reflects the market's belief that its dominance will allow the company to continue to generate growth in earnings at a pace vastly exceeding that of traditional industrial enterprises.

Dissecting analysis results by comparing ratios of key parameters is a powerful approach, and one to use frequently. Any time you have two numbers to compare, this kind of "ratio analysis" can lead to important insights into the causes of underlying differences between the two numbers.

TABLE 27.1. Comparison of financial statistics for Microsoft and General Electric

	Units	Microsoft	General Electric (GE)	Ratio Microsoft/GE
Annual revenues	B 1999 $	14	100	0.14
Number of shares	Billions	5	3.3	1.5
Price per share	1999$/share	87	112	0.77
Earnings per share (EPS)	1999$/share	1.17	2.88	0.41
Price-to-earnings (PE) ratio	Ratio	74	39	1.9
Earnings per dollar of revenues	$/dollar	$0.42	$0.10	4.2
Market capitalization	B 1999 $	433	371	1.2
Market cap. per dollar of revenues	$/dollar	$31	$3.7	8.3

(1) B signifies billions (thousand millions for British readers).
(2) PE is the share price to earnings ratio.
(3) Market capitalization is the product of # of shares and price per share. Price per share is the product of EPS and PE ratio.
(4) Values are current as of April 1999.

Creating a presentation or an executive summary for a complex report always involves boiling an analysis down to the essentials. First create equations (like the ones above) that calculate key analysis results as the product of several inputs multiplied together; then, determine which of these inputs affects the results most significantly. Creating such models can help you think systematically about your results.

28

APPLYING THESE LESSONS

Let's look at a specific example to make this advice more real. The web site <http://www.stateline.org> has time series data by state for a variety of items, including educational and environmental expenditures as well as tax revenues. As an exercise, I downloaded some data from the site and incorporated it into an annotated Excel spreadsheet (statedata.xls) so that readers could explore the various steps I took to clean, analyze, and understand the data. You can download this spreadsheet at <http://www.numbersintoknowledge.com>. The data in this example include population, personal income per capita, state tax collected, state expenditures on environmental agency budgets, and state expenditures on education.

I first saved each stateline.org HTML table as text, and imported it into

Excel.[62] These raw data then needed to be cleaned up a bit. I deleted blank rows, removed the Virgin Islands and Puerto Rico from the per capita income columns (because not all data series had values for these U.S. territories), moved the U.S. totals to the bottom, and checked each column of data to make sure that the order of the states was identical (sometimes there are slight differences in data order within such files; it's worth at least spot checking). I also scanned the data for numbers that were suspiciously small or large and found none. In addition, I added an "independent U.S. total" row, so I could add up the state data and make sure it matched against the U.S. totals reported in the data.

I then adjusted the data so that all aggregate dollar figures were expressed in millions of dollars. I reordered the columns to put population and personal income first because they are more fundamental data that could be used to normalize the other data in various ways. I calculated total personal income by multiplying per-capita income by population. I adjusted 1996 current dollars to 1997 dollars using the implicit GDP deflator. The 1996 education expenditures were not given in the stateline.org data, but 1997 expenditures were given along with the percentage growth rate from 1996 to 1997. I then calculated 1996 education expenditures from these data. After all this "cleaning," my spreadsheet contained the data shown in **Table 27.2**, but for every state, not just for the six states shown here.

At this point, and only then, was I ready to analyze the data. I'd corrected for inflation, and I'd checked to make sure that the data were not obviously flawed. I had also systematized my spreadsheet tables so that each of them started at the same row (for example, California would be on row 19 of all the spreadsheets). This technique makes it much easier to link to the spreadsheets and create calculations in an automatic way.

The first normalization was to calculate state tax collections, environmental agency budgets, and education expenditures on a per-capita basis (**Table 27.3**). On this basis, state environmental agency budgets are only tens of dollars per person per year; education budgets are generally about 50 times bigger per capita. Across the U.S., per-capita state tax collections amount to about 6.5% of total per capita personal income.

I also noticed that the education expenditures per capita were a very substantial fraction of total state tax collection per capita. In New Mexico, for example, these expenditures total more than 80% of all state taxes collected. This result seemed suspicious to me, so I investigated further.

TABLE 27.2. Cleaned data from stateline.org for selected states

State	Population Millions		Per-capita personal income 1997$/person/year		Total state tax collections M1997$/year		State Environment agency budgets M1997$/year		State Education expenditures M1997$/year	
	1996	1997	1996	1997	1996	1997	1996	1997	1996	1997
California	31.9	32.3	25,610	26,314	NA	61,667	677	686	35,498	35,546
New Mexico	1.7	1.7	18,981	19,298	NA	3,102	38	37	2,512	2,516
New York	18.1	18.1	29,555	30,250	NA	34,865	300	329	17,602	16,243
South Carolina	3.7	3.8	20,017	20,508	NA	5,381	169	227	3,898	3,903
Texas	19.1	19.4	22,761	23,707	NA	23,025	408	387	17,425	18,303
West Virginia	1.8	1.8	18,453	18,724	NA	2,906	235	196	2,114	2,386
Total U.S.	265.2	267.6	24,614	25,298	NA	443,493	7,020	7,481	268,423	275,821

(1) State tax collection data were not available for 1996.

TABLE 27.3. Per-capita (PC) data for selected states

State	Per-capita personal income 1997$/person/year		PC state tax collections 1997$/person/year		PC Environment agency budgets 1997$/person/year		PC Education expenditures 1997$/person/year	
	1996	1997	1996	1997	1996	1997	1996	1997
California	25,610	26,314	NA	1,911	21	21	1,114	1,102
New Mexico	18,981	19,298	NA	1,793	22	21	1,468	1,454
New York	29,555	30,250	NA	1,922	17	18	971	896
South Carolina	20,017	20,508	NA	1,431	45	60	1,049	1,038
Texas	22,761	23,707	NA	1,184	21	20	913	942
West Virginia	18,453	18,724	NA	1,600	129	108	1,161	1,314
Total U.S.	24,614	25,298	NA	1,657	26	28	1,012	1,031

(1) State tax collection data were not available for 1996.

9 I pulled out my *1997 Statistical Abstract of the U.S.*, and found Table 237 (School Expenditures, by Source of Funds in Constant [1993-94] Dollars: 1980 to 1994) and Table 477 (All Governments—Revenue, Expenditure, and Debt: 1980 to 1994). As usual, the *Statistical Abstract* tables give footnote references to the original sources, which would allow me to check the details if necessary.

The latest data from the *Statistical Abstract* were for 1994, but they were recent enough to serve as a check on the stateline.org data. Table 477 from the *Statistical Abstract* showed 1994 total state tax revenues of $373B in 1994$ ($396B in 1997$), and growth over the preceding four-year period of $70B, so I could easily imagine state tax collections reaching the $444B figure (in

1997$) from the stateline.org data by 1997. This calculation confirmed that the stateline.org data for tax collections were plausible. Taxes are not the only source of state revenue, however. Total state revenues include a category called "Current charges and Miscellaneous" in Table 477, and these additional revenues increase the revenues from state taxes by about 30%. To completely sort this out, I would have had to find the original source of these estimates and determine whether these "miscellaneous" revenues are used to support education in any way. I didn't search for this information in this case because of the illustrative nature of this example.

The education expenditures took a bit more investigation. One complexity is that some of the state expenditures on education come from money contributed by the Federal government (this money is not dependent on state tax receipts). Another complexity is that the state education expenditures include *local* government expenditures, but the state tax receipts do *not* include local tax revenues. Table 237 from the *Statistical Abstract* helped make this point crystal clear because the state expenditures on education in 1994 in this table total only $165B in 1994$ ($175B in 1997$), which is significantly lower than the $276B shown in Table 27.2. The local expenditures must account for the rest.

If the local tax revenues are added to the state tax receipts, the fraction of tax receipts represented by education would be much lower than it appears from Table 27.3. We solved the puzzle, but it took comparison with an independent source to do it.

Another way to normalize the data is as a fraction of the national average, as shown in **Table 27.4**. This approach shows quickly which states are higher or lower than the mean. In this sample, California and New York have per-capita personal incomes above the national average (their state tax collections per capita, as well as those of New Mexico, are also above the norm). South Carolina and West Virginia both show per-capita environmental agency budgets well above the U.S. average. In fact, these numbers are so much higher than the average that they probably warrant further investigation.

Per-capita state education expenditures in California, New Mexico, South Carolina, and West Virginia also exceed the national average, but, as we saw above, care must be used in interpreting these numbers. The federal government also contributes funds to education, and these funds are not always distributed equally on a per capita basis among the states.

Still another way to normalize the data is as a percentage of per-capita personal income, as shown in **Table 27.5**. This view of the data shows how impor-

**TABLE 27.4. Per-capita (PC) data normalized to the national average
for selected states**

State	Per-capita personal income Natl average = 1.0		PC state tax collections Natl average = 1.0		PC Environment agency budgets Natl average = 1.0		PC Education expenditures Natl average = 1.0	
	1996	1997	1996	1997	1996	1997	1996	1997
California	1.04	1.04	NA	1.15	0.80	0.76	1.10	1.07
New Mexico	0.77	0.76	NA	1.08	0.84	0.77	1.45	1.41
New York	1.20	1.20	NA	1.16	0.63	0.65	0.96	0.87
South Carolina	0.81	0.81	NA	0.86	1.71	2.16	1.04	1.01
Texas	0.92	0.94	NA	0.71	0.81	0.71	0.90	0.91
West Virginia	0.75	0.74	NA	0.97	4.88	3.85	1.15	1.28
Total U.S.	1.00	1.00	NA	1.00	1.00	1.00	1.00	1.00

(1) State tax collection data were not available for 1996.

**TABLE 27.5. Per-capita (PC) tax collection and expenditure data
as a percentage of per-capita personal income (PI) for selected states**

State	Per-capita personal income PC PI = 100%		PC state tax collections PC PI = 100%		PC Environment agency budgets PC PI = 100%		PC Education expenditures PC PI = 100%	
	1996	1997	1996	1997	1996	1997	1996	1997
California	100%	100%	NA	7.3%	0.1%	0.1%	4.4%	4.2%
New Mexico	100%	100%	NA	9.3%	0.1%	0.1%	7.7%	7.5%
New York	100%	100%	NA	6.4%	0.1%	0.1%	3.3%	3.0%
South Carolina	100%	100%	NA	7.0%	0.2%	0.3%	5.2%	5.1%
Texas	100%	100%	NA	5.0%	0.1%	0.1%	4.0%	4.0%
West Virginia	100%	100%	NA	8.5%	0.7%	0.6%	6.3%	7.0%
Total U.S.	100%	100%	NA	6.6%	0.1%	0.1%	4.1%	4.1%

(1) State tax collection data were not available for 1996.

tant each of these items is compared to an important measure of the total
wealth of our society.

Finally, I calculated percentage growth from 1996 to 1997 in total and per-
capita data, as shown in **Table 27.6**. For time-series data over a longer time
period, it is also useful to calculate an index relative to the base year, as shown
in Figure 27.4 (above). New York and West Virginia show comparable growth
rates between total and per capita data; the other states and the U.S. average

TABLE 27.6. Percentage growth 1996–97 in total and per-capita data for selected states

State	Population	Total				Per Capita			
		Personal Income	State Tax collections	Environ-mental agency budgets	Education expen-ditures	Personal income	State tax collections	Environ-ment agency budgets	Education expen-ditures
California	1.3%	4.1%	NA	1.4%	0.1%	2.7%	NA	0.1%	-1.1%
New Mexico	1.1%	2.8%	NA	-2.2%	0.1%	1.7%	NA	-3.3%	-0.9%
New York	0.0%	2.4%	NA	9.6%	-7.7%	2.4%	NA	9.6%	-7.7%
South Carolina	1.2%	3.7%	NA	34.9%	0.1%	2.5%	NA	33.3%	-1.0%
Texas	1.8%	6.1%	NA	-5.2%	5.0%	4.2%	NA	-6.9%	3.2%
West Virginia	-0.3%	1.2%	NA	-16.8%	12.9%	1.5%	NA	-16.6%	13.2%
Total U.S.	0.9%	3.7%	NA	6.6%	2.8%	2.8%	NA	5.6%	1.8%

(1) State tax collection data were not available for 1996.

show lower growth rates (or faster declines) on a per-capita basis than for the aggregate totals. Of course, two sequential years of time series data are not terribly illuminating, but the same techniques can be applied when analyzing longer time series.

CONCLUSIONS

When John Holdren was a professor at UC Berkeley, he taught a delightful class titled Tricks of the Trade. In this class he described many of the unwritten rules about being effective in the energy/environment field and listed key pitfalls in data acquisition and handling. I aggregate them below into four golden rules:

- *Avoid data that are mislabeled, ambiguous, badly documented, or otherwise of unclear pedigree.* Ambiguity and poor documentation are an indication that the quality control for such data is uneven at best and appalling at worst. Dig into the numbers a bit and find out whether it's carelessness or incompetence; make sure you believe the numbers before using them.

- *Discard unreliable data that are invented, cooked, or incompetently cre-*

33

ated. If you find major inconsistencies, conceptual flaws, and omissions in the data, you'll need to discard them, no matter how much they might help your analysis.

- *Beware of illusory precision.* Don't represent or interpret data as more accurate than they are. Carefully characterize uncertainty and variability in your data, and insist that others do so with their own.

- *Avoid spurious comparability.* Beware of numbers that are ostensibly comparable but are fundamentally inconsistent. Create and use only consistent comparisons.

25

Holdren's advice when dealing with data is to "be suspicious, skeptical, and cynical. Assume nothing." Though it may sound paranoid to the uninitiated, such caution is an absolute necessity for the seasoned problem solver.

For every foolproof system, there is a bigger fool.

— **EDWARD TELLER**

28

MAKE A MODEL

In this book, the term "model" refers to any conceptual (usually mathematical) model that can be used to understand the world. Models can range in scope from the back of an envelope to millions of lines of computer code. Anthony Starfield et al., in their book *How to Model It*, summarize nicely when they state "An explicit model is a laboratory for the imagination." Models allow you to experiment, test, and learn about the world in a systematic way without leaving your office.

SIMPLE MODELS AND BURIED ASSUMPTIONS

A model is an equation or set of equations that describe how something you care about changes when other things change. Below is a useful model that allows you to estimate gasoline consumption as a function of how many miles you drive:

$$\text{Gas used per week (gallons)} = \frac{\text{Miles driven}}{\text{Week}} \times \frac{\text{Gallon}}{\text{Miles}}$$

You can estimate your car's average fuel efficiency based on past experience in a typical week. If you drive 100 miles per week, and your car consumes five gallons of gasoline in that period, your car gets 20 miles per gallon (mpg) on average. For every additional 100 miles you drive, you will consume another five gallons of gasoline.

This model is perfectly adequate for many purposes, but a more detailed model may be required in some cases. Say that you want to estimate your gas consumption for a special week-long, 1,000 mile trip that will be much more

on the highway than your normal pattern. The following model would then be useful:

$$\text{Gas used} = \frac{\text{City miles driven}}{\text{Week}} \times \frac{\text{Gallon}}{\text{City miles}} + \frac{\text{Highway miles driven}}{\text{Week}} \times \frac{\text{Gallon}}{\text{Highway miles}}$$

Let's say that the car's efficiency is 15 mpg in the city and 25 mpg on the highway, and that your big trip is 95% highway driving, as shown in **Table 28.1** (about 67% of the driving is on the highway in the typical driving pattern embodied in the simpler model). The more complex model indicates that the vehicle would consume a total of 41.3 gallons, while the simpler model (which is based on your typical driving pattern) yields an estimate of 50 gallons when applied to your special trip. The simpler model therefore overestimates gasoline use by 21% in this case.

There are many other subtleties that could be incorporated into such a model, which would be introduced as additional terms in the equation. Neither one of these models is precisely correct in a scientific sense although the second one is more generally applicable. Either can be appropriate in certain situations.

The assumption about how much city and highway driving you do is *buried* in the first model. It's easy to see how someone not aware of the difference in automobile efficiency in city and highway driving could misuse the first model and calculate an erroneous result. This lesson is a more general one about models: assumptions are often buried and can lead you astray.

SIMPLIFY, SIMPLIFY, SIMPLIFY

People are often intimidated by the use of computer models to conduct an analysis, but never forget that even computer models depend on data and assumptions that are based in large part on analysts' judgment. It is a rare model that does not contain dozens or hundreds of such assumptions; in fact, these data and assumptions determine the outcome in many cases. This heavy reliance on assumptions for parameters that are unknown is most prevalent in models used in business and public policy, but even models of the physical world can rely on assumptions that influence or determine the outcome.

TABLE 28.1. Simpler and more complex models of automobile fuel use

	Driving Mode	Simpler model	More complex model	Ratio of Simpler to More complex
Miles per gallon	City		15	
	Highway		25	
	Average	20	24.2	0.83
Miles driven	City		50	
	Highway		950	
	Total	1000	1000	1.00
Estimated gallons consumed	City		3.3	
	Highway		38	
	Total	50	41.3	1.21

In his real estate practice, Robert J. Ringer encountered sellers with fantastic expectations for the investment value of their buildings:

> As a general but very consistent rule, owners are so unrealistic when it comes to vacancy factors, replacement costs, and other expense items that are not readily ascertainable that I normally considered their projections and operating statements to be virtually meaningless.[63]

Using his electronic calculator and his knowledge of real estate, Ringer was able to determine quickly whether the building was salable at or near the owner's asking price. His simple conceptual model allowed him to create a quick estimate of net cash flow on the proverbial back of an envelope. This screening technique allowed him to avoid wasting time on deals that would never close because of the huge gap between the reality of the project's net cash flow and the fantasy of the owner's expectations.

The technique of using a simple and transparent model to evaluate the results of a more complex calculation is immensely powerful. Analysts at the International Institute for Applied Systems Analysis (IIASA) found this out to their chagrin when Will Keepin, a Visiting Scholar at IIASA, was able to almost exactly reproduce the results of a multiyear, multimillion dollar study using some of the study's key input assumptions and a hand calculator. Keepin

29

showed that the study's results followed directly from the input assumptions. He concluded that the study's projections of future energy supply "are opinion, rather than credible scientific analysis, and they therefore cannot be relied upon by policy makers seeking a genuine understanding of the energy choices for tomorrow."

Beware of big complicated models and the results they produce. Generally they involve so much work to keep them current that not enough time is spent **26** on data compilation and scenario analysis. Morgan and Henrion, in their book *Uncertainty*, devoted an entire chapter to such models and began by summarizing this fundamental truth:

> There are some models, especially some science and engineering models, that are large or complex because they need to be. But many more are large or complex because their authors gave too little thought to why and how they were being built and how they would be used.[64]

Such large models are essential only for the most complex and esoteric analyses, and a simpler model will usually serve as well (and be more understandable to your intended audience).

The importance of transparency cannot be overestimated. A model that your audience can actually grasp is inherently more persuasive than a "black box" that no one outside of a small circle of analysts understands. Transparent mod- **33** els for which the input data and assumptions are also well documented are even more compelling, but are, sadly, all too rare.

Don't be too impressed by a model's complexity. Instead, ask about the data **26** and assumptions used to create scenarios. Focus on the coherence of the scenarios and their relevance to your decisions, and ignore the marketing doublespeak of those whose obsession with tools outweighs their concern with useful results. Sadly, many modelers fall into this latter camp, and you would do well to avoid them.

Programming today is a race between software engineers trying to build bigger and better idiot-proof programs, and the universe trying to produce bigger and better idiots. So far, the universe is winning.

— RICH COOK

REUSE OLD ENVELOPES

Ross and Ross, Mattimore, and von Baeyer describe how the physicist Enrico Fermi used to dole out exam problems but not supply all the information necessary for a solution. His students (my former professor Art Rosenfeld was among them) were expected to make educated guesses about the missing parameters to solve these so-called "Fermi problems." This technique is not often taught in school, but it is one well worth learning. For virtually any problem, it is possible to create an approximate solution using information you know or can estimate from daily life experience. Most people don't believe this until they try it a few times, but it's true.

FROM OIL TO WATER

I read an article once that described how Global Water Corp. of Vancouver had signed a contract with an Alaskan city (Sitka) that would allow the company to harvest water from glacial runoff.[65] The company was going to take converted oil tankers and ship the water to China, where it would then be bottled and sold. I initially thought that this venture was absurd, but a few back-of-the-envelope calculations made it clear that there was method to this apparent madness.

The company signed a fifteen-year contract covering 4.8 billion gallons of water a year. The article claimed that half-liter bottles of water from Europe currently cost Chinese consumers about $5.25 while this new source of water would cost about $0.70 per half liter bottle ($1.40 per liter).

The oil tanker was originally meant to carry crude oil, which, at the time, was selling for about $20 per barrel. There are 42 gallons per barrel of oil, yielding a cost per gallon of about $0.50 (gasoline typically sells for twice that amount in the U.S. because of the additional refining and transportation costs

to bring the fuel to gas stations, as well as federal and state taxes). There are about 4 liters per gallon (a liter is a bit more than a quart), so the price of oil per liter is about $0.12.

The bottled water would (assuming the article was correct) sell for $1.40 per liter, as described above, which is about ten times greater than the price of the original cargo for the tanker (oil). This calculation showed that the company could make some serious money shipping water to China, assuming that the market for bottled water was large enough to sell the tanker's contents. Of course, there are costs of cleaning and converting the tanker to hold pure water, and costs of packaging the bottled water that a gasoline vendor would not incur, but the costs of transport would be comparable. The one-time cost of converting the tanker would be relatively small when amortized over many trips. I guessed that the costs of packaging and marketing the water would at most amount to between ten and fifty cents per liter. These costs would reduce the profits of the venture, but it still seemed likely to be profitable after these adjustments.[66]

MORE CALCULATIONS

I once sat outside an air-conditioned store in Cambridge, Massachusetts that had its door propped open on a very hot day. Using some simple assumptions and conversion factors I had memorized, I calculated how many dollars per hour the store was wasting by refrigerating the outside world. That calculation was enough to convince the store manager to close the door (he probably never had anyone badger him about *that* before, even with all the crazy scientists wandering around Cambridge!).

Another calculation I did about ten years ago was to figure out the wattage of a candle. All I needed to assume was that wax is similar to petroleum (I had the energy content of petroleum memorized) and note the time it took for the candle to burn away. I also needed to know some conversions to turn English units into metric units. It turns out that typical dinner candles dissipate energy at a rate comparable to that of a 60 to 100 watt light bulb (of course, candles are much less efficient in creating light than are electric lamps, so they don't create nearly as much light as a bulb).

USING THE BACKS OF ENVELOPES

Such "back-of-the-envelope" calculations, supplemented by what Mattimore calls "smart guessing," are not just for scientists. A business plan for a new venture, for example, usually involves dozens of such smart guesses about sales of a product, costs of sales, overhead, and other parameters. The first draft of such a plan is essentially a back-of-the-envelope calculation that will later be refined and improved in discussions with experienced business people.

People involved in fast-moving new ventures usually don't have time to research issues thoroughly and don't have money to hire someone to do it for them (sometimes even the people they hire don't have time either!). Often, back-of-the-envelope calculations are the best that can be done in the time allotted. The law of averages keeps the accuracy of such calculations from deteriorating too much because you are just as likely to overestimate as to underestimate (assuming that you're not biased one way or another). Across all the assumptions you make, these deviations average out.

EXERCISE

Estimate how many cobblers there are in the U.S., using just information you have in your head. Check your answer and method against that used in the first chapter of John Harte's book *Consider a Spherical Cow*.

THINKING ABOUT THE PROCESS

There are many ways to create these quick calculations, but the first step is always to create a one-equation *model* by breaking the problem into its component parts. This equation might look like the relationship between market capitalization and earnings per share in the chapter titled "Dig into the numbers":

$$\text{Market Capitalization (\$)} = \text{\# Shares} \times \frac{\text{Earnings}}{\text{Share}} \times \frac{\text{Price}}{\text{Earnings}}$$

It might also look like the model of automobile fuel economy shown in the chapter titled "Make a model":

$$\text{Gas used} = \frac{\text{City miles driven}}{\text{Week}} \times \frac{\text{Gallon}}{\text{City miles}} + \frac{\text{Highway miles driven}}{\text{Week}} \times \frac{\text{Gallon}}{\text{Highway miles}}$$

An important part of this process is balancing accuracy against the time needed to create the calculation. Back-of-the-envelope calculations are meant to be quick, so they usually rely on relatively simple models.

Once the relationships between key inputs and the desired output are established, you'll need to determine which of the inputs you know and which you'll have to assume for purposes of the initial calculations. *Don't get hung up on a particular number.* The best thing to do is to carry out the calculation using the very roughest of approximations for those things you don't know, and refine the inputs later.

BOUNDING THE PROBLEM

When you've set up your calculation with your best guesses for the various input parameters, it's often instructive to identify those that are least certain and to determine a plausible range for them. You then carry through the calculation using the upper and lower ends of the ranges for all the inputs.

For example, if you only knew the fuel economy of your car very roughly, you might specify a range of 12 to 18 miles per gallon in the city and 22 to 30 miles per gallon on the highway for the "gas used" equation above. By combining the lower ends of both efficiency ranges in your calculation, you could determine an "upper bound" to how much gasoline you might use on the trip. This calculation could be useful in making sure that you don't ever run out of gas along the way or that you have enough money to pay for gas in your budget. Using the upper ends of the efficiency ranges would give you the minimum cost for gas that is likely for the trip.

These kinds of bounding calculations actually represent simple scenario analyses. If you compute ranges for the inputs and the results don't vary much from your best estimate, you've learned that the calculation is insensitive to the uncertainties you've identified, which is an important result. There are, of

The desire for quantification can go too far.

DILBERT *reprinted by permission of United Feature Syndicate, Inc.*

course, much more sophisticated ways of characterizing uncertainty, but for the purposes of the simple calculations discussed here, bounding the ranges for key parameters is often good enough.

APPROXIMATIONS

Most people think of numbers as immutable, but when doing rough calculations, it's perfectly OK to fiddle with the numbers a bit to make the calculations easier. For example, say you need to divide 3,500 by 40. Instead of reaching for your calculator, you can say "40 is only a bit more than 35, and 3,500/35 = 100. Therefore, 3,500/40 is a bit less than 100." These approximations can save you time when only a rough answer is needed. Don't be afraid to make these simplifications when you need to.

UNIT ANALYSIS

Another simple but powerful tool in back-of-the-envelope calculations is called "unit analysis." Think of the calculation we did above to determine the cost of oil in $/liter. When creating the model, you could just list the numbers and multiply them together, like this:

$$\text{Cost (\$/liter)} = \frac{20}{(42 \times 4)} = 0.12$$

Alternatively, you could list each of the terms with units attached so that it is obvious that they cancel out and yield the end result in units we desire:

$$\text{Cost (\$/liter)} = \frac{\$20}{\text{barrel}} \times \frac{1 \text{ barrel}}{42 \text{ gallons}} \times \frac{1 \text{ gallon}}{4 \text{ liters}} = \frac{\$0.12}{\text{liter}}$$

I initially scoffed when my high school chemistry teacher taught this methodical way of carrying through and checking our calculations. At the time, the calculations were so simple that I didn't need to be methodical and could just do them in my head. I've learned through experience, however, that calculations are rarely that simple in practice, and, even if they are simple, I still make mistakes because I'm tired or rushed. It pays to write out the units for your calculation to make sure you are calculating correctly.

OTHER HINTS

To become truly proficient at these techniques, memorize key conversion factors, at least in a rough way. Which ones you memorize depend on your chosen field. Energy and environmental conversion factors are most useful in my work, but I have also memorized many other conversions because they often come in handy.

If you simply can't memorize such items, then create a small page with common conversion factors and keep it in your wallet. Graduate students at the Energy and Resources Group at UC Berkeley are routinely supplied with several of these pages during their course of study, and one of these is reprinted in "Facts at your fingertips." While creating such a "cheat sheet" may brand you as a nerd in the eyes of your compadres, it will surely impress them when you can answer questions while only consulting your data sheet.

One modern tool that is invaluable for such calculations is the spreadsheet (proving that you don't need an envelope to create back-of-the-envelope calculations). Create quick calculations using a spreadsheet and plug in guesses for those parameters you don't know at the time. This approach allows you to think through the calculations without hindrance, and then go back later to check the input assumptions. Later modifications to these assumptions trickle through the calculations nicely without requiring much additional work.

CONCLUSIONS

Hans Christian von Baeyer points out that doing such calculations instead of relying on authority engenders self-confidence and independence. Achieving even occasional success in back-of-the-envelope estimation increases your inclination to tackle such problems in the future, thereby ensuring that you will gain further experience and self-assurance. To know in your heart that you can roughly estimate just about anything is a marvelous feeling of mastery.

EXERCISE

Find three quantitative assertions in today's newspaper, and do back-of-the-envelope calculations to check them. Did you find anything surprising?

Whether the problem concerns cooking, automobile repair, or personal relationships, there are two basic types of responses: the fainthearted turn to authority—to reference books, bosses, expert consultants, physicians, ministers—while the independent of mind delve into that private store of common sense and limited factual knowledge that everyone carries, make reasonable assumptions, and derive their own, admittedly approximate, solutions.

— HANS CHRISTIAN VON BAEYER

USE FORECASTS WITH CARE

The future is uncertain, but people keep trying to forecast it anyway. In an article with an ironically humorous and utterly extraneous subtitle– "Few Wall Street Analysts said that '96 would be this good: Forecasts for '97 may or may not be just as inaccurate"–Chet Currier of the Associated Press assessed the sorry record of most financial analysts in predicting the huge increase in stock prices in 1996.[67]

After describing how virtually all analysts predicted modest gains or little change from end-of-1995 levels, Currier states that the major market indices racked up gains of 20% or more in 1996. His conclusions about the value of such forecasts are particularly insightful:

> None of the errant prognostications cited above were wild guesses or inter-pretations of the leaf patterns in some hallucinogenic tea. They were the carefully considered projections of respected commentators who enjoy wide followings as financial analysts. All of them remain in their same jobs, or comparable positions of prominence, a year later, and are presumably busy right now preparing to make their projections for another year.
>
> As usual, these commentaries may well provide a useful point of depar-ture for your own investment planning in 1997. *The reasoning behind a pre-diction is often the most interesting and useful thing about it* (italics added). But the experience of this year dramatizes how skeptically you may take all forecasts of something as unpredictable as the stock market – even from people who know what they are talking about.

Newspapers always trumpet the results of such forecasts, but, as the stock mar-ket example shows, you should be wary.

Steven Schnaars, in his book *Megamistakes: Forecasting and the Myth of Rapid Technological Change*, gives many examples of forecasts of the popu-

larity of new products that were totally off the mark. In fact, about 80% of the forecasts he examines were grossly inaccurate, which made him question the entire basis for predicting adoption of new technologies. He concluded, after reviewing dozens of examples, that explicit scenario analysis is far superior to most conventional ways of exploring the future.

26

Forecasts are used

- for planning (either institutional or personal);
- for advocacy (to lend credence or context to concern over a particular issue);
- for research (to assess the potential effects of current or expected future trends).

Most institutions create forecasts for their own budget planning. Such a forecast attempts to capture recent trends and allows the institution to plan for expected growth or declines in revenues. A typical forecast in this context would look one or two years ahead, sometimes three or four years in exceptional cases. Longer-term forecasts are used for government institutions (like Social Security) that have obligations spread over many years. Forecasts can also serve as part of a scenario-planning exercise.

An institution concerned with a particular issue will sometimes create a forecast that validates concern for that issue, as part of its effort to create free publicity. One edition of *The Sunday San Francisco Chronicle/Examiner* contained two separate examples where a forecast was cited for this purpose.[68] In one case, a California real estate investor and analyst predicted that housing prices in California would double over the next eight years. In another, the Cremation Association of America predicted that the rate of cremation, measured as the % of deaths that lead to cremation, would increase from 21% in 1997 to 38% in 2010. In both cases, a person or institution developed forecasts that showed why a particular issue of concern should also be of interest to the public.

The investor who predicted a doubling of California home prices was roundly criticized by his colleagues, but he succeeded in getting the attention of the press, in part because his prognostication conflicted so glaringly with the conventional wisdom. That strategy is a successful one that is widely used: predict something just a tad outrageous, and use your credibility from previous

work to lend credence to your prediction; then talk to as many reporters as will listen in hopes they will mention you and your company in their columns.

More complex issues can require detailed forecasts to truly understand them, which usually involves computer modeling. In these applications, researchers will create a baseline forecast that attempts to predict how the world will evolve assuming current trends continue. This forecast is often called a "business-as-usual" case. Then, one or more alternative forecasts are created, to assess the potential effects of changes in key factors on the results. For example, an economic forecaster might use such an analysis to assess the likely effects of a change in property taxes on economic growth in a particular state.

THE WONDERS OF EXPONENTIAL GROWTH

One of the most basic concepts in forecasting is that of exponential growth. This concept generally applies to population forecasts because the rate of population growth is proportional to the number of people who are alive at any time (the more people there are, the more babies are born). Similarly, money in a bank account grows exponentially because the interest earned is related to how much money is in the account at any time.

Forecasts of population or economic activity are often summarized in the form of annual exponential growth rates. For example, the annual growth in gross domestic product was roughly 3.5% in 1996. If this growth continued at 3.5% per year for five years, total growth would be 1.035 x 1.035 x 1.035 x 1.035 x 1.035 - 1, or 19%.

Exponential growth can lead to surprising results in the longer term. **Table 30.1** compares three countries of 100 million people, one of which (Country A) grows two million people per year every year, and another of which (Country B) grows 2% every year (two million in the first year). After one hundred years, Country A will have grown to 300 million people, while Country B will have grown to more than 700 million! The differences become even more pronounced for higher growth rates. If Country C grew at 4%/year, its population would grow to more than five billion people a hundred years hence. A doubling of the exponential growth rate leads to a sevenfold increase in total population!

The famous futurist Julian Simon once made the statement that "We have the technology to house, feed, and clothe a growing population for six billion

TABLE 30.1. Exponential growth and population

	Units	Country A	Country B	Country C
Annual growth	Millions of people/yr or %/yr	2	2%	4%
Initial population	Millions	100	100	100
Population in 100 years	Millions	300	724	5050
Total growth over 100 years	Millions	200	624	4950
Average growth per year	Millions/year	2	6	50

years."[69] When told that any reasonable population growth rate would lead to the mass of all people exceeding the mass of the known universe over that period, he revised his statement to six *million* years. Even this revision wasn't enough to get Simon out of trouble, though. In six million years at modest growth rates, the number of people would exceed the number of electrons in the known universe!

Forecasts projecting exponential growth can lead to trouble when the rate of growth changes dramatically. During the 1950s and 1960s, growth in U.S. electricity use increased at a roughly constant 7%/year. Power planners had an easy job until the 1970s when the oil crisis, increasing energy efficiency, and problems with large power plants led to price increases and slowing demand for power. Growth rates in electricity demand slowed to 2-3% per year, and utilities were forced to cancel the planned construction of hundreds of power plants because of the drastically reduced growth in power use.

A KEY TRICK OF THE TRADE

In evaluating forecasts of exponential growth rates, an important trick is known as "the rule of 70." If you have an annual growth rate (say 2% per year for U.S. population) then the time it will take (in years) for population to double is roughly $\frac{70}{2}$ or thirty-five years. Growth of 7% per year implies a doubling time of about ten years. You don't need to know why this works, you just need to remember it.

Why is a doubling time useful? It is much easier to calculate multiples of two

than to actually calculate exponential growth in the formal way. When only an approximate number is needed, a doubling time is good enough. Say you want an approximate estimate of how much $10,000 now will be worth when you retire. If the growth rate after inflation is 7%/year (a reasonable long-term estimate for money invested in a tax-deferred stock account) the doubling time is ten years. In 30 years, the money will have undergone three doublings, or a factor of eight increase (2 x 2 x 2), and will be worth about $80,000 in 1997 dollars (of course, you'll have to pay taxes on that money when you withdraw it).

This approximation becomes less accurate as the growth rate increases. For rates above 10% per year, you'll probably want to check your results using more accurate methods, but for most situations, the rule of 70 is more than good enough (see *Ecoscience* for more details).

THE LOGISTIC CURVE

Many phenomena in nature and in business are described by the so-called "logistic curve" (also known as the "S curve"). This curve (shown in **Figure 30.1**) is characterized by slow but rising exponential growth in the early phase, rapid growth in the middle phase, and growth that tapers off as fundamental limits are reached (see *Ecoscience*). The logistic curve describes most phenomena eventually and applies equally well to technology adoption and population growth. We don't know where the top limits are in many cases, but in cases where we can identify those upper limits, the concept can come in handy.

One such example relates to the number of Internet users, which, at one point, was doubling every 100 days. At this rate, it would reach four to eight billion people by 2002 or 2003, which is roughly the number of people in the entire world. This upper limit is a firm one that cannot be exceeded (barring the unlikely discovery of an alien world that wants to hook up to our computer networks). We can therefore infer that this growth will start leveling off over the next few years. The system may grow in other ways, of course, but, in this one aspect, the limits will not be exceeded. Knowing this fact can help a company that is planning its Internet business strategy cope with the future. The company can be sure that these upper limits will become important relatively soon, and, if its business model depends on continuous growth in numbers of users, the company will need to modify it.

FIGURE 30.1. A logistic curve

COPING WITH AN UNCERTAIN FUTURE

Forecasts are an integral part of the Cycle of Action. They are most useful in developing the action sequence, but they can also be important in the process of goal setting and in the evaluation of how actual events compare to our goals. In spite of their failings, and whether they are created using a formal model or developed in an intuitive way, they are an essential part of life and work.

If forecasts are part of your planning process, don't create just one. Instead, use a set of forecasts (i.e., scenarios) to explore the future. Vary key factors, and investigate which of them to ignore and which to dissect further. All forecasts are wrong in some respect, but if the process of designing them teaches you something about the world and how events may unfold, creating them will have been worth the effort.

Finally, it is important to adopt strategies that are robust in the face of inevitably imperfect forecasts. Several computer companies have moved to "build-to-order" manufacturing, which allows them to assemble computers as requested by customers. This strategy reduces dependence on forecasts but introduces other challenges in manufacturing (which are surmountable using current technology). This same lesson applies equally well to other such decisions: if the key variables are difficult or impossible to foresee, then use scenario analysis to evaluate the possible outcomes, and adopt strategies that are less dependent on forecasts.

EXERCISE

Find a forecast on a topic you care about, and learn how it was created. Talk to the author and read any supporting documentation. Do you still find it plausible?

In the space of 176 years, the Lower Mississippi has shortened itself 242 miles. That is an average of a trifle over one mile and a third per year. Therefore, any calm person who is not blind or idiotic can see that in the Old Oolitic Silurian Period, just a million years ago next November, the Lower Mississippi was upward of 1,300,000 miles long. By the same token, any person can see that 742 years from now, the Lower Mississippi will be only a mile and three-quarters long. There is something fascinating about science. One gets such wholesale returns of conjecture out of such a trifling investment of fact.

— **MARK TWAIN**

HEAR ALL SIDES

One of the first rules of management in dealing with a conflict between coworkers is to listen carefully to all sides of the story before saying or doing *anything*. The same lesson applies in evaluating any two courses of action or points of view. It almost always pays to hear eloquent supporters of divergent viewpoints address each other's criticisms in a discussion or debate format. If you do not include such discussions as regular fodder for your deliberations, you are missing out on one of the most potent ways to clarify any set of issues. In rare cases, of course, such debates can muddy the waters still further, but they are almost always helpful as long as the participants are well intentioned, well matched, and prepared.

President John F. Kennedy knew this when he created a cadre of advisors who held widely divergent viewpoints during the Cuban Missile crisis. After the Bay of Pigs debacle, Kennedy vowed never to rely solely on his military advisors again. He created an ad hoc group of advisors, led by his brother Robert, to bring to his attention other alternatives. One of these options was the blockade of Cuba, which was ultimately the strategy that Kennedy adopted.

In the academic world, there is a tradition of "brown bag lunches" where researchers present work-in-progress to friends and close colleagues. These lunches help identify problems with analysis before a paper is written or presented to a wider group. They work best when all concerned are on friendly terms and the objective of clarifying the key argument is accepted by all participants.

The business world does not generally encourage the kind of interaction found in such "brown bags," to its detriment.[70] One acquaintance of mine works for a large mutual fund company. She lamented to me that the people under her supervision did not possess the ability or inclination to check their own work, which often led them to produce analysis that did not answer the management questions at hand. These analysts were responsible for creating reports that the firm used to assess financial fitness, so the reports were impor-

tant. Unfortunately, her analysts had not learned what constituted good analysis, in part because the firm had no tradition of constructive intellectual criticism.

If you have reached a conclusion, it still pays to have a critic review your work. In fact, such review is virtually always a good idea (even at early stages of your analysis). Confirm your results using as many independent sources as you can find. These can be only approximate comparisons, but you should always do this, just like you would test computer equipment before you put it in the field.

14

It is a capital mistake to theorize before you have all the evidence. It biases the judgment. — **SHERLOCK HOLMES**

SHOW YOUR STUFF

Don't forget the final step in any analysis: making your results useful to others. Your reports should be treasured by your colleagues as classics full of pertinent data, exhaustive documentation, and clear thinking. If they're not, you should follow the advice contained in these chapters and change your ways.

If you follow this advice, you'll soon gain a reputation for analytical integrity and clear thinking that can't be beat. If you don't, you'll be contributing to the piles of mediocre and barely used reports that inhabit bookshelves everywhere, and you most certainly don't want to do that.

They say that 90% of television is garbage. But 90% of EVERYTHING is garbage. — **GENE RODDENBERRY**

KNOW YOUR AUDIENCE

How results are expressed is at least as important as what the results are. Most analysts are so involved in their calculations that they forget that other people don't care nearly as much about the results of the analysis as they themselves do. Good analysis is often handicapped in its presentation because the authors have not taken the vital last step of thinking about what the audience cares about and expressing the results in those terms.

One of my favorite professors from graduate school, Art Rosenfeld, is a master at making complex analysis results understandable to the lay person. Art presented his estimates of energy savings from the latest efficient gadget (a new refrigerator, for example) for many different people. Sometimes he would testify before congress, other times he would address public utility commissioners, and still other times he would tell his story for a general audience. He would always tailor his talk to the listeners. If oil drilling in the Arctic National Wildlife Refuge was a hot topic before congress that day, he would express his energy savings estimates in terms of "Arctic Refuges saved." If the public utility commissioners were concerned about canceling a power plant construction project, Art would calculate how much the savings were worth in terms of "avoided power plants." And for the lay audience, Art would show how much money people would save per household if they bought the better fridge. In each case, Art's message gained resonance because he thought seriously about the concerns of the audience.

In both talks and papers, your audience will only retain a few key points from your work. *You* must decide what those few points should be, and hammer them home. If you don't, you risk not reaching the people you set out to convince. So step back and *reflect* after you've completed your analysis. Think **21** about what you've learned, and explain it in three or four simple sentences. *Dig into the numbers* to support your main points, but don't overwhelm your audi- **27** ence with numbers. Instead, focus on the conclusions, describe why your

analysis makes such a compelling case, and explain it in ways specifically tailored to reach your audience.

This lesson is particularly important when the analysis results are intended to affect public opinion or a public policy debate. Journalists generally latch on to only a few key results of a complex study when deciding what "angle" to take with a story. Express these results to reach out to the desired audience, and your effectiveness will increase many times over.

In order to achieve victory, you must place yourself in your opponent's skin. If you don't understand yourself, you will lose one hundred percent of the time. If you understand yourself you will win fifty percent of the time. If you understand yourself and your opponent you will win one hundred percent of the time. — **TSUTOMU OSHIMA**

33

DOCUMENT, DOCUMENT, DOCUMENT

Over the years, friends have clipped dozens of articles for me on topics in which they knew I had an interest. In almost all cases, they neglected to write on the clipping the full reference, thus rendering it unusable for research purposes. An article without a reference is useless except as background info. An article with a reference is a citable source.

Documentation lends *credibility* to any intellectual effort. It also recognizes the previous work of others and allows readers to follow your thought processes (it also allows YOU to recreate your thinking months after you have achieved some conceptual breakthrough). Any competent analyst ought to be able to recreate your analysis from the documentation you provide, and you must be able to do the same more quickly than others can. Finally, the process of documenting your results can help you check those results and ensure accuracy.

GIVING CREDIT WHERE CREDIT IS DUE[71]

Progress in intellectual endeavors depends on prior work being fairly acknowledged. When you have relied on a particular fact or insight that you got from another source, you must reference it. You can put this reference in a footnote, you can use the (Author Date) method (e.g., (Koomey 1995)), or you can consult *The Chicago Manual of Style* for more reference formats.

If you quote from another source, cite the source (and go back to the original source if you can). If it is an extensive quotation, if the quotation represents the "heart" of the work, or if using the quotation might affect the commercial value of the cited work, it may be appropriate to obtain the author's permission for using it. "Fair use" has a specific legal definition, but there are many gray areas. You are allowed to quote excerpts from other people's copyrighted work to illustrate, clarify, or comment on their work. Whether an excerpt constitutes

fair use depends on the context in which the excerpt is used and on a variety of other factors. One common-sense test to apply is to ask yourself "If the cited source were my work, how would I feel if someone else cited it in the way I cite it here?"

For more details on copyrights in the digital age, see the Copyright Office web site for the Library of Congress <http://www.copyright.gov>, and the Nolo.com site devoted to this topic <http://www.nolo.com/>. For information on obtaining permission for the use of copyright protected material, see Richard Stim's excellent book titled *Getting Permission*, also available from Nolo.com.

The Internet has made issues of credit and authorship more complex. One friend of mine occasionally teaches a class on environmental science, and he found that in term papers submitted by his students, three-quarters of all the sources cited were World Wide Web addresses (called Uniform Resource Locators, or URLs), with little or no explanation given for how the information was derived or whether it was credible. For the students, the mere existence of a URL was enough reason to cite it when in fact more investigation and documentation is really required to ensure that the web site gives proper credit and accurately reports results.

If you do rely on a web site for a research project, don't forget to print out a hard copy of the site. Web sites change all the time, and their impermanence places special responsibility on the researcher to record the information in a permanent way. Another important detail is to set your web browser to include the actual URL in the header or footer of any printed copies of web sites. Some browsers do this automatically, but yours may not, so make sure this setting is saved as the default option.

LEAVING A TRAIL FOR YOU TO REMEMBER

The fancy technical name for data documentation is "source tagging." Researchers who specialize in data quality focus on how to decide which information about a source is appropriate for electronic data and how to attach this key information to those data. I refer to data without attached sources as "disembodied." I also call them useless, because without the sources, they are little better than rumors.

Don't underestimate the importance of good archiving practice: Archiving is the process by which older data and records are stored for retrieval if needed. Companies neglect this process at their peril because mission-critical information must be accessible to those making strategic decisions. This task takes on greater importance in the electronic age because the "paper trail" that existed previously is now commonly absent (see Cook's 1995 *Technology Review* article for details).

Keep a notebook for each major project: Record in this book important insights you have, notes from meetings, reports to track down, or any other items related to the progress of your work. My electronic organizer serves this function, but if you haven't adopted one of these handheld devices, then a paper notebook will serve just fine. These notes may come in handy later if you want to build on previous work.

Always date your work, even your roughest handwritten notes: Reconstructing the evolution of your thoughts often requires knowing *when* you wrote something down.

Always write down the full reference to an article that you've copied, as soon as you've copied it: Write it on the first page of the copied article. No exceptions! Until I learned this lesson, I can't tell you how many times I had to go back and find the source because I had not written down the reference at the moment I copied the article.

Always write down why you think a particular article is significant for your work, because in six months you probably will have forgotten: Any longer-term research project involves more details than you can keep in your head, so keep good notes on articles you decide to clip.

Be selective about the articles you clip and file, but clip the important ones without fail: In this age of ubiquitous web access, electronic databases of most publications are widely available. Save time and avoid clipping articles unless they are immediately useful for a current research project. Do print out and file those articles you actually use, so you can refer to them later.

If you create computer code or spreadsheets, put extensive explanations of your logic into the code in comments and footnotes: Describe your reasoning as if someone else will be reading that code and trying to understand it because that is literally the case: in six months, you will be a very different person who has forgotten the details that led you to structure your calculations in the way you did.

If you work with spreadsheets, never type data directly into spreadsheet formulas: For example, some people enter formulas that look like this into a spreadsheet cell: = 2 x 457.3/9415 + 24. Instead, enter these numbers into cells, then create formulas that use these cells for calculations, like this: = F2 x F3/F4 + F5 (in this case, the various numbers are in cells F2 through F5). You can easily change the data later without having to fiddle with the formulas, plus you can see the input data for the calculation without having to poke around in the cell containing the formula. Such planning ahead can save you time and make it easier to spot mistakes in the data and calculations.

Segregate input data in one part of the spreadsheet and results in another part: Be systematic and you will be able to recreate your calculations even a year from now, after you've forgotten the assumptions that are now so fresh in your mind.

Define units: Make sure your documentation is clear about exactly which units you are using. For example, the word "ton" has at least five different possible meanings. There are short tons (2,000 pounds or 907 kg), long or deadweight tons (2,240 pounds or 1,016 kg), metric tons (sensibly defined as 1,000 kg), tons of cooling (defined as 12,000 Btus/hour of cooling capacity) and register tons, which are (oddly enough) defined in units of volume! Also beware of definitions that vary by country. The word "billion" means 1,000,000,000 (10^9) in the U.S., but in much of Europe that same number is written as "thousand million." In Europe, a billion is 1,000,000,000,000 or 10^{12} (which in the U.S. is called a "trillion"). For an excellent on-line dictionary of units, see <http://www.unc. edu/~rowlett/units>.

Define constants: Constants and conventions must be carefully defined, or confusion can result. For example, analysts typically assume that a year is equal to 8,760 hours, which is 365 days times twenty-four hours per day. In reality, the average year is 8,766 hours, because of leap years (every fourth year has an

additional day). The small difference in these two numbers may lead to slight discrepancies, but other such conventions may have larger effects.

Go metric: Use metric units in any analysis reports that may be useful to readers outside the U.S. If your U.S. audience would find those units confusing, include *both* English and metric units. Because the World Wide Web is an international medium, you should ensure that your work will gain wide use by those in other countries who use different (and in this case more logical) conventions than you do. Tailoring your results for the ease of these readers will help your message spread more quickly than it would otherwise.

`38`

`32`

CREATING A TRAIL FOR OTHERS TO FOLLOW

The importance of documenting your work comes to the fore when there is some kind of a crisis. In August 1997 my boss, Mark Levine, was called before the President's Council of Economic Advisors to testify about a report we had recently completed. I had created most of the calculations summarizing the results and had written a detailed description of how the calculations were done. I crafted the documentation for a technical analyst who wanted to reproduce my calculations and who had time to delve into depth. While Mark is a technical analyst, it soon became obvious that he needed something a little different because his time was so limited. I talked him through the calculations, and the testimony went well. The lesson was not lost on me, however.

Assume that the person reading your documentation has limited time to figure it out. This is the right approach when documenting because it will ensure that you take appropriate care. You may think that your task won't require such thoroughness, but what if someone else has to pick up where you left off?

Describe the calculations in such excruciating detail that, even in the worst case, an intelligent reader could understand and reproduce your calculations as quickly as possible. Put a clear description of your methods in the main body of your report; then, relegate the complete documentation to appendices. Such detail may seem like overkill at the time, but you will never regret it. You may not be able to anticipate how your calculations will be used in the future. Documentation is one of Stephen Covey's "important but not urgent" tasks and one that should be a priority in any analysis work.

`10`

"I think you should be more explicit here in step two."

© 2000 Sidney Harris from cartoonbank.com. All Rights Reserved.

Your references to other people's work must be complete so that others can follow up. References must contain enough information to enable a reader to find a reference at the library even in the face of the inevitable typos and smudged copies. Redundancy is good in this case. **Table 33.1** shows the information you should include for the most commonly referenced sources. How you choose to present this information in your bibliography is up to you, but I recommend following the basic formats in the *Chicago Manual of Style* (a nice summary of this and other reference formats is contained in *A Writer's Reference*, by Diana Hacker).

Computer programs that automate the process of tracking references (such as EndNote) are one of the few innovations since the word processor to increase significantly the productivity of writers and researchers. Use them if you do more than a little writing. These tools can ensure that your references are complete and do so in a fraction of the time and annoyance that in the past accompanied compiling a bibliography. These tools allow you to create one database of your references, and, once you've entered them into the database, you need never enter them again. It formats the references in any way you choose and

TABLE 33.1. Information needed for complete references

Type	Author	Year	Title	Institution	Volume	Number	Page #	Date	URL	Other
Book	Yes	Yes	Yes	Publisher	Edition		If applicable		Yes	City
Edited Book	Yes	Yes	Yes	Publisher	Edition		If applicable		Yes	City
Book Section	Yes	Yes	Yes	Publisher	Edition		Yes		Yes	Book title, City
Journal Article	Yes	Yes	Yes	Journal name	Yes	Yes	Yes	Yes	Yes	
Magazine Article	Yes	Yes	Yes	Magazine name	Yes	Yes	Yes	Yes	Yes	
Newspaper Article	Yes	Yes	Yes	Newspaper name			Yes	Yes	Yes	Section, City
Report	Yes	Yes	Yes	Name of institution			If applicable	Yes	Yes	City
Conference proceedings	Yes	Yes	Yes	Conference organizer	Yes		Yes	Yes	Yes	Location
Thesis	Yes	Yes	Yes	University			If applicable	Yes	Yes	Thesis type, University location
Personal Communication	Yes	Yes	Yes	Name of institution			If applicable	Yes	Yes	Address, phone #, fax #, email

comes with a collection of formats for many of the more common academic journals. Acquire reference database software with all due haste!

CHECKING YOUR WORK

You wouldn't put a computer system into a client's office without first testing it, would you? Why then do so many people produce analysis without checking the results for internal consistency, completeness, and congruence with the stated goals of the people requesting the analysis? An important function of the documentation process is to help conduct this testing. Explaining your methods in detail helps identify conceptual blunders. If you can't explain the reasoning behind your methods, then you have more thinking to do.

Documentation should be an integral part of your analysis process. As you are conducting your research, you should write down your methods or try to explain them to a colleague. Your analyses will be the better for this give and take.

22

IF YOU ARE WORKING WITH DOLLARS

The value of money is affected by inflation. Over time, inflation makes each dollar worth less, so a dollar spent in 1997 is *not* the same as a dollar spent in 1990. You must, therefore, document the year in which expenditures occur.

Any time you present tables or figures that include dollars, you *must* specify "current dollars" or "constant dollars in some year." Current dollars imply that if the money is spent in 1990 it is in 1990 dollars and if spent in 1997 it is in 1997 dollars. Alternatively (and more accurately in my view), you can correct for inflation using an inflation index from the *Statistical Abstract of the U.S.* to calculate all expenditures in constant dollars. The most common indices for this purpose are the Consumer Price Index and the implicit Gross Domestic Product (GDP) deflator (sometimes called the "chain-type price index for GDP"), but sometimes a more specialized index for a particular material or industrial sector comes in handy. The expenditures in current dollars in 1990 would then be converted to 1997 dollars by multiplying them by the following factor:

$$\frac{\text{Inflation index in 1997}}{\text{Inflation index in 1990}}$$

If the inflation index in 1997 is 1.5, and the inflation index in 1990 is 1.2, then expenditures in 1990 must be multiplied by $\frac{1.5}{1.2}$ or 1.25 to convert them to 1997 dollars. All other years would be treated similarly using the appropriate inflation index (of course, 1997 expenditures are already in 1997 dollars).

Correcting for inflation, although simple, is commonly overlooked. When the U.S. Postal Service announced in May 1998 that it would be granted an increase in the price of a first-class stamp in 1999, *The San Francisco Chronicle* presented a summary of the new rates and printed a graph of first-class postal rates since 1885.[72] This graph showed a huge increase in prices, from 2¢ in 1885 to 33¢ in 1999. An uninformed observer might easily have concluded from this graph that the price of first-class mail had skyrocketed, particularly since the 1970s. The problem, of course, is that the graph was not corrected for inflation, and nowhere did the *Chronicle* article mention this omission. *The Oakland Tribune* story on the same topic[73] correctly reproduced two graphs, one in current dollars and one in inflation-adjusted dollars (these graphs are reproduced as **Figures 33.1** and **33.2**). The second graph showed that the 33¢ price of first

FIGURES 33.1 and 33.2. U.S. postal rates in nominal (current) and real dollars

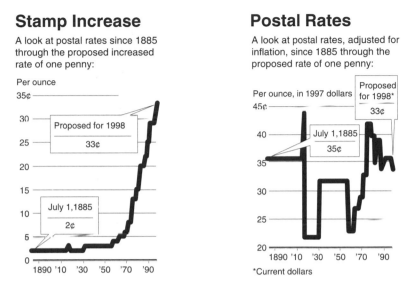

Reprinted with permission of AP/Wide World Photos.

class mail in 1999 actually represented a decline in constant dollars from 35¢ in 1885.

I once found a report that contained information on the costs of computing power in the 1950s and 1960s, and at first I was delighted. My delight turned to dismay when I realized that the report did not document whether the costs were current or constant dollars, and I was unable to use the information because I couldn't determine how to treat the costs. What a waste! Don't let this happen to your work.

There are similar (but even more intricate) subtleties regarding the use of currency exchange rates to convert foreign currencies to dollars or vice versa. Because such rates fluctuate over time, you must document which exchange rates you are using; otherwise your calculations are useless to the reader.[74]

TAKE RESPONSIBILITY FOR YOUR WORK

When documenting or writing reports, take responsibility by using the active voice. For example, say "I estimated the following parameters" instead of "the following parameters were estimated." Make it crystal clear who did what, so there is no ambiguity. The passive voice is the tool of people who want to avoid

From *The Big Book of Hell* © 1990 by Matt Groening. All Rights Reserved.
Reprinted by permission of Pantheon Books, a division of Random House, Inc., NY.

responsibility for their actions, but the purpose of documentation is to assign responsibility, for all the reasons listed above. Don't thwart your efforts by using the often ambiguous passive voice.

DOCUMENTATION PAYS OFF

Any time there is a new "hot" analysis topic, people with little prior knowledge jump in. You can perform an immense service to any debate (and garner atten-

tion for your point of view) simply by being systematic in identifying and investigating such topics and documenting them in a careful way.

A case in point is the raucous debate over the cost of pollution to society from power plants during the late 1980s and early 1990s. In 1990, I wrote a paper that was not a conceptual breakthrough but simply created consistent comparisons among different estimates of these social costs. I have received hundreds of requests from all over the world for this admittedly rather dry technical report as well as offers to speak at conferences and opportunities to contribute to other reports on the topic. It garnered attention in large part because of its systematic and careful documentation. You too can use this approach to your advantage.

The same sort of intellectual chaos that pervaded that debate now characterizes the emerging demand for advertising and commerce on the web. Businesses want hard numbers about the effectiveness of web advertising in selling their products, but solid statistics are scarce. Systematic thinking in this area could be of immense value to the companies deciding on advertising budgets. As of this writing, the decisions are being driven more by hype than data.

BACK UP YOUR DATA!

Many analysts don't routinely back up their electronic files. This oversight is potentially disastrous, as the otherwise technology-savvy Stanford Business School found out in April 1998. Two large computers were moved before the operators verified that all data had been backed up, and almost a dozen researchers lost years of work.[75] Hard disks, floppy disks, and data tapes don't last forever, and they are guaranteed to fail at the least opportune time (they can also become obsolete as technology changes). I have many colleagues who have lost data from such failures, and it's never fun. So after you've gone to the trouble of carefully documenting your work, make sure that there are at least a couple of electronic copies, and put a paper version in a safe place. One trick is to email electronic files to yourself, which places them on the server of your Internet service provider and adds an additional layer of redundancy. Also don't forget that data formats change over time, so if you have some older files, convert them to a new format before trashing the program that created them.

CONCLUSIONS

Ignorance of good data documentation practices is widespread. These skills should be taught in high school, yet technically trained college graduates routinely do not identify their data sources or the methods they used for their calculations. Don't fall into this trap. Make sure that any intelligent person (including YOU!) can reproduce your analysis from the footnotes. With this one simple step, you will vault above 95% of the world's analysts. Leaping over the other 5% depends on the quality of your content.

Results! Why, man, I have gotten a lot of results. I know several thousand things that won't work. — **THOMAS A. EDISON**

34

LET THE TABLES AND GRAPHS DO THE WORK

Don't just write to be understood; write so that you cannot be mis-understood. — **ROBERT LOUIS STEVENSON**

In any writing that relies on quantitative analysis, the calculations are the bulk of the work. First, create the tables documenting data and presenting results; then, write the text describing the inputs and analysis. For particularly important results, make graphs that best emphasize the points you want to make.

36

35

The text should call out only the most important numbers in the table for ease of comprehension. The numbers junkies can delve into the table for any details they feel compelled to unearth.

The following passage is an example of the right way to summarize technical analysis, taken from the section "Dig into the numbers" above:

27

> Table 34.1 summarizes the key financial parameters for calculating market capitalization for the two companies.
>
> The first thing to notice about the financial statistics for these two companies is a huge disparity: Microsoft's market capitalization in April 1999 was about 20% larger than that of GE in spite of revenues that were only one-seventh as large. All other things being equal, we might expect that companies with similar market valuations would also have similar revenue streams. All other things are *not* equal, however, and finding out why will help illustrate this important analytical technique.
>
> If Microsoft's market capitalization per dollar of revenues was the same as that of GE, we'd expect that its market capitalization would be only one-eighth as large as it is. We need to explain this eightfold discrepancy, and, to do so, we examine the components of the last equation above. The first

TABLE 34.1. Comparison of financial statistics for Microsoft and General Electric

	Units	Microsoft	General Electric (GE)	Ratio Microsoft/GE
Annual revenues	B 1999 $	14	100	0.14
Number of shares	Billions	5	3.3	1.5
Price per share	1999$/share	87	112	0.77
Earnings per share (EPS)	1999$/share	1.17	2.88	0.41
Price-to-earnings (PE) ratio	Ratio	74	39	1.9
Earnings per dollar of revenues	$/dollar	$0.42	$0.10	4.2
Market capitalization	B 1999 $	433	371	1.2
Market cap. per dollar of revenues	$/dollar	$31	$3.7	8.3

(1) B signifies billions (10^9).
(2) PE is the share price to earnings ratio.
(3) Market capitalization is the product of # of shares and price per share. Price per share is the product of EPS and PE ratio.
(4) Values are current as of April 1999.

component is earnings per dollar of revenues, and the second component is the price-to-earnings ratio.

As Table 34.1 shows, Microsoft's earnings are about four times greater than GE's per dollar of revenues, which accounts for about half of our eight-fold difference. The price-to-earnings ratio also differs between the two companies. Apparently, the stock market values one dollar of Microsoft's earnings twice as much as one dollar of GE's earnings, which accounts for the remaining factor of two.

Now compare it to the text below that doesn't follow the approach I advocate:

Table 34.1 summarizes the key financial parameters for calculating market capitalization for the two companies.

The first thing to notice about the financial statistics for these two companies is a huge disparity: Microsoft's market capitalization in April 1999 ($433B) was about 20% larger than that of GE ($371B) in spite of revenues that were only one-seventh as large ($14B vs. $100B). All other things being equal, we might expect that companies with similar market valuations would also have similar revenue streams. All other things are *not* equal, however, and finding out why will help illustrate this important analytical technique.

If Microsoft's market capitalization per dollar of revenues ($31) was the same as that of GE ($3.7), we'd expect that its market capitalization would be only one-eighth as large as it is. We need to explain this eightfold discrepancy, and to do so, we examine the components of the last equation above. The first component is earnings per dollar of revenues, and the second component is the price-to-earnings ratio.

As Table 34.1 shows, Microsoft has 5B shares outstanding, with earnings of $1.17 per share. GE has about 3.3B shares in the hands of stockholders, with earnings of $2.88 per share. The product of the number of shares and the earnings per share gives the total annual profits for each company. For Microsoft, profits are about $6B/year on revenues of $14B, and for GE, profits are about $9.5B on revenues of $100B.

Every dollar of Microsoft's revenues therefore yields four times more profits than a dollar of GE's revenues ($0.43 compared to $0.10 per dollar of revenues). So the increased profitability of Microsoft per dollar of revenues accounts for about half of our eightfold difference.

The price-to-earnings ratio also differs between the two companies. Apparently, the stock market values one dollar of Microsoft's earnings (PE = 74) twice as much as one dollar of GE's earnings (PE = 39), which accounts for the remaining factor of two.

This text is potentially intimidating to the reader who is less comfortable with numbers, it is less compact, and it repeats data that are already presented in the table itself.

In addition, if the calculations change, you'll have to manually change the text to reflect those calculations. If the calculations are in a table printed or copied from a spreadsheet, you merely change the calculations in the table, print it out again or copy it into your text, and change the few key numbers in the text that refer to the table. This approach is far less labor intensive when you need to make changes (and you almost always need to make changes at some point).

If you omit the table and just include the more numbers-heavy text, as is commonly done, you make it even harder for readers to assess the validity of your assertions. The table organizes the information and makes it conveniently available to the reader. It also makes it easier to write the text.

The main purpose of writing of this type is to present, in as clear a fashion as possible, the logic of the calculations, the main input data, and the key

33 results. Any competent analyst ought to be able to reproduce your analysis using the documentation contained in your report. If they can't, your work isn't done. An astounding number of analysts routinely omit vital data and assumptions from their reports, a practice for which there can be no excuse.

If the analysis is well conceived, the tables and graphs well designed, and the audience clearly defined, the report should practically write itself. I normally use the following main headings in technical reports (your specific needs may dictate a slightly different structure):

> Introduction
> Methodology and Data
> Results
> Discussion
> Limitations of this analysis
> Future work
> Conclusions
> References
> Appendices

Using this method, the writing task becomes one of straightforward description, except in the "Discussion" section, where you describe why the results are important.

Place detailed calculations and supporting data in appendices. Only put *essential* information in the main body of the report. When possible, write relatively short technical reports that summarize the analysis results and relegate complete documentation to appendices. It is then a simple matter to take the short report, eliminate all but the three to five of the most crucial graphs and tables, change some text, and send it off to a journal for publication.

One trick that helps the reader of longer articles and reports is to put in bold face the first mention of tables and graphs in the text so it is obvious where they are discussed and explained. For example: "**Table 1** shows the results...." This trick makes it easier for readers who jump straight to the results and want to comprehend the analysis quickly. There are many such readers, which also **33** makes it essential that each table and figure have footnotes that document the key elements of the supporting calculations.

There may occasionally be good reasons for breaking the rules described above. However, by following them most of the time, and by thinking carefully about the concerns of your intended audience, you will most effectively convey the implications of your results.

32

The design of the data structures is the central decision in the creation of a [computer] program. Once the data structures are laid out, the algorithms tend to fall into place, and the coding is comparatively easy.

— **BRIAN W. KERNIGHAN AND ROB PIKE**

35

Modern graphs are largely an eighteenth-century creation. Edward Tufte, the modern master of graphical display, states that Playfair, Lambert, and others developed the first graphs that broke away from a geographical representation. In the two centuries since, such charts have become commonplace. With the advent of modern computer tools, creating graphs from data involves trivial effort. In fact, it has probably become too easy. Graphs are often produced without thought for their main purpose: to enlighten and inform the reader.

Tufte defines graphical excellence as "that which gives to the viewer the greatest number of ideas in the shortest time with the least ink in the smallest space."[76] He promotes several rules that will help you avoid the most common pitfalls, which I summarize below:

- *"Above all else show the data"*: Make sure the graphs highlight your results instead of hiding them as many graphs do. Use the graphs to tell stories about your analysis, and make those stories leap out at the reader.
- *"Maximize the data-to-ink ratio, within reason"*: Tufte defines the data-ink ratio as the amount of ink that conveys information divided by the total amount of ink used for the graph. Many graphs have low data-ink ratios, and you should avoid this pitfall. To maximize the data-ink ratio, "erase non-data-ink and redundant data-ink."
- *"Eliminate chart junk"*: Remove clashing fill patterns (Moiré patterns), which are hard on the eyes. Eliminate extraneous gridlines, which obscure the data. Delete graphical elements that decorate but convey no information, such as 3-dimensional (3-D) chart elements on 2-D graphs.

 This latter mistake is a common one, particularly in "slick" brochures and company annual reports. Although there are legitimate uses for 3-D

graphs, a bar chart is not one of them. I adapted **Figure 35.1** from the 1999 annual report of a company with $6 billion in sales in 1999.

FIGURE 35.1. The wrong way to make a bar chart

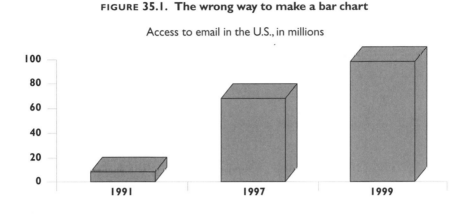

Figure 35.1 shows access to email in the U.S. using 3-D bars. Because of the 3-D perspective, it is impossible to extract the data accurately from the graph. **Figure 35.2** shows how the graph should look for greatest accuracy.

FIGURE 35.2. A better way to make a bar chart

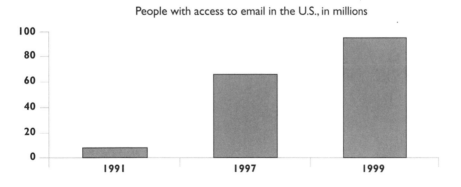

- *"Maximize data density" (numbers per unit area), within reason:* Typical graphs in newspapers and journals, based on Tufte's review of hundreds of charts published 1970-1980, have data densities of 5 to 20 numbers per square inch (0.8 to 3 numbers per square centimeter). The densest graph Tufte examined (a map of galaxies contributed by some astronomers) had a data density of 110,000 numbers per square inch (17,000 numbers per

square centimeter); the least dense graph he considered had a data density of 0.15 numbers per square inch (0.02 numbers per square centimeter). Maps tend to have high data density; badly designed charts tend to have the lowest.

When things get really bad, a graph may contain only one data point. I encountered such an appalling graph on an airplane flight from San Francisco to Washington, DC on June 8, 1999. It showed the size of the market for an electronic gadget, and it looked like **Figure 35.3**:

FIGURE 35.3. The market for an electronic gadget

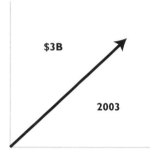

Figure 35.3 was all that appeared on a 17" diagonal TV screen, which represents a data density of 0.008 numbers per square inch (0.001 numbers per square centimeter), probably an all-time low. It gives no indication of the current size of the market except that the arrow going up and to the right implies that the market is growing from today until 2003. Avoid this blunder and maximize data density.

- *"Use small multiples"*: One important way to maximize data density is to create what Tufte calls small multiples, defined as "a series of graphics, showing the same combination of variables, indexed by changes in another variable."[77] Let's say you plot monthly rainfall for Mobile, Alabama, as shown in **Figure 35.4**. If you have data on rainfall for many cities that you'd like to compare, you can shrink them and plot them side by side, as shown in **Figure 35.5**. You need not reproduce all the details of the chart labels on every chart. Once the reader has learned the "key" to reading the chart, small multiples like these allow quick and easy comparison of large amounts of data.

- *"Revise and edit"*: All work improves with revisions. Leave enough time to get feedback and make your graphs better.

FIGURE 35.4. **Average monthly rainfall for Mobile, Alabama, 1961-1990 (inches)**

Source: *Statistical Abstract of the U.S. 1997*, Table 397.
Normal Monthly and Annual Precipitation–Selected Cities (based on 1961-1990 data).

FIGURE 35.5. **Average monthly rainfall for selected cities, 1961-1990 (inches)**

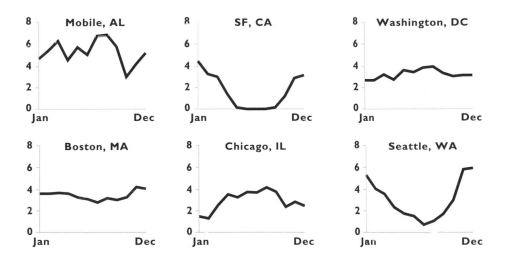

I add one more rule of my own, which is *Make figures stand alone*. Figures are often copied separately from the report in which they appear, to allow the reader to trace your figure back to the source. Put the date and file name on the figure, at least when it's in draft form. Make sure the final version contains some reference to the report in which it appears or to the author's contact information. Figures are often copied in black and white even if the original graph is in color. Make sure a black and white copy conveys the same message as the original, and give the reader the report and author's contact information on the graph as a backup in case the colors don't come through in the reader's copy.

FIGURE 35.6. A graph that needs work

Shipments of fluorescent ballasts for the U.S. commercial sector

NOTE: Because they are not typically used in the commercial sector, the following ballast types are omitted from the data presented in this figure: low-power-factor ballasts (shipments of 24.2 million in 1996), 1500 MA high-power-factor ballasts (0.2 million in 1996), and ballasts included in the "All other corrected power factor" category of the Census Bureau data. Ballast imports are not included because they were not disaggregated by ballast type in the census data. Arrows indicate the years in which state standards and, finally, federal standards took effect, preventing the sale of standard magnetic ballasts.

SOURCES: Koomey et al. (1995); Census Bureau MQ36C(94)-5, Table 3; Census Bureau MQ36C(96)-5, Tables 3 and 4

APPLYING THESE RULES

Let's look at a figure I created in the past and make it better using these rules. **Figure 35.6** shows a graph of the historical annual sales of fluorescent lamp ballasts for U.S. commercial buildings that appeared in a 1997 report.[78]

Notice that the patterns used in the area chart seem to "vibrate" when you stare at them. This is what Tufte calls the Moiré effect, and it's best avoided. The

FIGURE 35.7. An improved graph

Shipments of fluorescent ballasts for the U.S. commercial sector

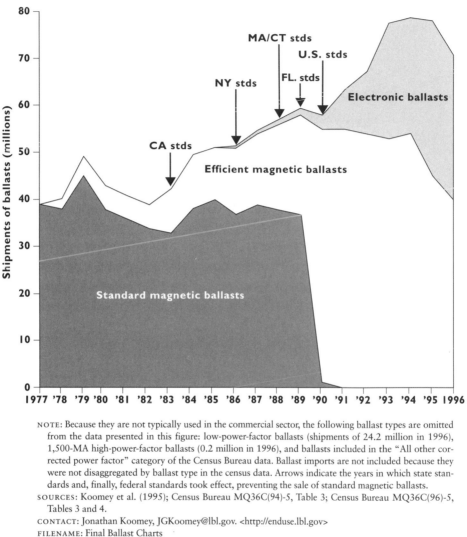

NOTE: Because they are not typically used in the commercial sector, the following ballast types are omitted from the data presented in this figure: low-power-factor ballasts (shipments of 24.2 million in 1996), 1,500-MA high-power-factor ballasts (0.2 million in 1996), and ballasts included in the "All other corrected power factor" category of the Census Bureau data. Ballast imports are not included because they were not disaggregated by ballast type in the census data. Arrows indicate the years in which state standards and, finally, federal standards took effect, preventing the sale of standard magnetic ballasts.
SOURCES: Koomey et al. (1995); Census Bureau MQ36C(94)-5, Table 3; Census Bureau MQ36C(96)-5, Tables 3 and 4.
CONTACT: Jonathan Koomey, JGKoomey@lbl.gov. <http://enduse.lbl.gov>
FILENAME: Final Ballast Charts
DATE OF CREATION: February 1995.
DATE OF LAST MODIFICATION: October 1997.

gridlines are extraneous, and the legend can easily be eliminated. The years along the X-axis are a bit hard to read because they are turned 90 degrees from horizontal. The footnote is descriptive and complete, but the figure still lacks the contact information for the author of the report and the file name of the graph.

Figure 35.7 is the improved version of the graph, which has a higher data-to-ink ratio than does Figure 35.6 and is easier to read. I eliminated the Moiré

vibration by using shades of gray and white and eliminated the legend by using labels directly on top of the appropriate areas of the graph. I removed the gridlines as well as the box around the graph and reduced the number of tick marks on the Y-axis to only the major ones. I shrank the type size of the years along the X-axis and shifted to a different convention of representing the years ('94 instead of 1994) for all years but the first and last (another choice would have been to lay out the graph in landscape format so that the X-axis was longer). Finally, I added author's contact information, the file name, and the dates of creation and last modification for the file.

CONCLUSIONS

Tufte reminds us that rules should not be followed slavishly and that our focus should be on the key goal: "to give visual access to the subtle and the difficult."[79] Graphics should serve your arguments. If they don't support your analysis in a direct and understandable way, they should be ruthlessly deleted or drastically modified. As Tufte says, graphs and figures that don't contribute to the "revelation of the complex"[80] have no place in your work.

Read Tufte's books. Review his guidelines periodically, and apply them to your current work. Small changes in graphics design can vastly improve the effectiveness of your figures and graphs.

Visual representations of evidence should be governed by principles of reasoning about quantitative evidence. For information displays, design reasoning must correspond to scientific reasoning. Clear and precise seeing becomes as one with clear and precise thinking.

— EDWARD TUFTE

36

CREATE GOOD TABLES

Agood table is a work of art while a bad one is worse than useless. I have worked to refine my table-making skills over the years, and I am constantly amazed at the perfectly dreadful tables created by authors who should know better. Tables are the heart of a technical report, and care in their creation is one sign of a clear-thinking analyst. A little care can make your tables a resource that your readers keep as a reference for many years.

I follow several key principles in creating a table:

- *Make them stand alone*: Tables are often copied separately from the report from which they came. Make sure the reader can trace your table back to the source (or at least to the author) and that your calculations are explained clearly in the footnotes.

- *Create extensive footnotes*: This aspect of documentation is crucial. **33** Footnotes to the table should explain the methodology, list the key sources, and describe how each source was used. Any intelligent person should be able to recreate the table from your footnotes. Tufte says that footnotes indicate "care and craft," and readers will be grateful when you create them.

 Of course, for tables appearing in more popular publications (like newspapers) the standards for footnote documentation are often lower, but even if the footnotes don't appear in the printed version, the person who created the table should still have the footnotes filed away in case there's a need to check the calculations later.

- *"Maximize the data-ink ratio," within reason*: Edward Tufte's guidance is **35** just as relevant for tables as for graphs. Make sure that every part of the table conveys information. Avoid extraneous gridlines (only include them when they enhance reader comprehension). Use carefully chosen white space to separate and group results.

- *Some redundancy is good*: Unlike in graphs, where redundancy should generally be avoided, showing additional columns or rows for totals and subtotals is essential. These ostensibly redundant sums help the reader follow the calculation and assure that the table is internally consistent. The sums can also help in troubleshooting your own calculations.

 Another important redundancy relates to percentages. If you have a column of percentages, label the column heading "% of total" AND put a percent symbol in every cell (e.g., 1.37%, not 1.37). Sometimes people make mistakes in their calculations and forget to divide by 100 to make a real percentage. By enforcing this redundancy, you guarantee that readers will interpret your results correctly and you'll save them time. Having the % symbol in every cell makes the meaning unambiguous.

 Finally, all numbers between zero and one should have a leading zero (e.g., 0.35, NOT .35). Copies become smudged and decimal points can be obscured. With a leading zero, there will never be ambiguity about the size of the number in question.

- *Enhance readability with font, style, and orientation choices*: In her *Non-Designer's Design Book*, Robin Williams demonstrates the principles of proximity, alignment, repetition, and contrast, and applies them to everyday design problems. Use spacing, justification (center, left, or right), grid lines, italics, and bold-facing to implement these principles within the table. Align decimal points within each column to add even more information.

 Choose a font that is easy on the eyes and doesn't deteriorate as a document is copied. Tufte recommends using the font used in telephone books (Bell Centennial) whenever you have tables that have high information density because the phone companies have researched this topic extensively, and there's no point in reinventing the wheel.

- *Enhance reader comprehension*: The single most important tool to increase usefulness of tabular data is to add rows or columns containing *indices* relative to a total or subtotal, which will show the relative importance of specific numbers relative to these totals (see the examples below). Most readers will calculate such indices themselves in their effort to understand the table. Do it for them, and they will thank you for it.

- *Normalize data to facilitate comparisons:* It's often helpful to normalize results on a per capita or per GDP basis, so the reader can compare them to more familiar quantities.

- *Create the table first; then, write your text from the tables*: Creating the **34** tables helps you think through the analysis and ensures that your logic is both correct and reproducible. Ask yourself questions about the analysis and check your work. If you can't explain it in writing, then you don't understand it. Clear tables and clear writing come from clear thinking.

- *Avoid spurious precision*: Only show the appropriate number of significant figures. Adding extra decimal places is distracting and pernicious. For example, results of calculations that use inputs reliable to one or two significant figures (e.g., 100 divided by 51) should not then be reported to five decimal places (1.93453). The correct value to use in this case is 2. The calculation is only as accurate as the least accurate input parameter (for more details on treating significant figures, see <http://www.chem.sc.edu/faculty/morgan/resources/sigfigs>).

- *Revise and edit*: Show your tables to colleagues before publication, and you will surely learn something. What is clear and obvious to you may not be so obvious to them. Give yourself time to revise and improve the tables.

A few specific examples will help illustrate some of these principles (A Microsoft Excel file [improvingtables.xls] containing the original and improved tables can be downloaded at <http://www.numbersintoknowledge.com>). The first is taken from a book written by a physics professor with many publications to his credit, and the publisher was one of the largest in the U.S.[81] **Table 36.1** shows how the original table appeared.

The first thing to notice is that the footnote is ambiguous. Is the total paper component split into 10 to 15% newspapers and 40 to 35% packaging, or is total paper split into 5 to 7.5% newspapers (50% times 10% to 15%) and 45 to 42.5% packaging? The former interpretation is more likely, so that's what I assumed for purposes of this example, but it is impossible to know for sure.

Table 36.2 shows what the improved table looks like. To make it, I first added a "Total" row, to remind readers that the total adds to 100%. I added gridlines to separate the main sections of the table, italicized the column headings, and changed the column headings to have lower case letters after the first letter of each word. I added the % symbol after each number to eliminate any ambiguity (this change was less important here than it would be if some of the values were less than 1). I incorporated the information contained in the original footnote into the table itself by showing subtotals under the "paper" cate-

TABLE 36.1. The table as it appeared in the book

Contents of American Landfills

MATERIAL	PERCENT OF VOLUME OCCUPIED
Paper	50*
Plastic	10
Organic (includes yard waste)	13
Metal	6
Glass	1
Miscellaneous (tires, rubber, debris)	20

*10 to 15 percent of the total volume is newspapers; the rest is packaging.

TABLE 36.2. An improved table

Contents of American Landfills[1]

Material	Percent of Landfill Volume Occupied	
Total paper	50%	
Newspaper		10 to 15%
Paper packaging		40 to 35%
Plastic	10%	
Organic[2]	13%	
Metal	6%	
Glass	1%	
Miscellaneous[3]	20%	
Total	100%	

[1]SOURCE: William Rathje, University of Arizona.
[2]Organic waste includes yard waste.
[3]Miscellaneous includes tires, rubber, and debris.

**TABLE 36.3. A table in one of my group's reports
that needs revision**

U.S. Shipments and Imports of Lamps, 1983-1994

Year	U.S. Shipments (millions of lamps)	Imports (millions of lamps)
1983	3615.9	560.9
1984	3723.4	748.7
1985	3472	862.7
1986	3421.3	920.6
1987	3399.4	999.8
1988	3510.2	1130.8
1989	3429.5	1024
1990	3318.5	1051
1991	3297.5	Data unavailable from Census Bureau
1992	3422.1	Data unavailable from Census Bureau
1993	3564.3	1372.6
1994	3563.3	1577.8

NOTE: "U.S. shipments" refers to total shipments by manufacturers located within the U.S., including units to be exported. Cold-cathode fluorescent lamps are excluded from the U.S. shipment data; Christmas tree lights are excluded from U.S. shipments as well as imports.
SOURCE: Census Bureau current industrial reports MQ36B, various years.

gory for newspaper and packaging. I right-justified these subtotals and reduced the type size to 8 points, to make it clear that these numbers are subordinate to the paper category and different than the other numbers in the table. I added a footnote giving the name of the source for the information and moved some of the qualifications for two of the waste categories into two other footnotes. This last change makes the additional information accessible to the interested reader, but removes it from the main table where it distracts the general reader from quickly grasping the main results.

The second example, shown in **Table 36.3** is taken from one of my group's technical reports.[82] This table slipped through our quality control, but I'm making amends by dissecting it for this example. Note the extraneous gridlines and the lack of indices that would help the reader focus on the key results from the table.

TABLE 36.4. Revised and improved version of Table 36.3

U.S. Shipments and Imports of Lamps, 1983-1994

Year	U.S. Shipments		Imports		
	Millions of lamps	Index 1983 = 1.00	Millions of lamps	Index 1983 = 1.00	Imports as a % of U.S. shipments
1983	3,616	1.00	561	1.00	16%
1984	3,723	1.03	749	1.33	20%
1985	3,472	0.96	863	1.54	25%
1986	3,421	0.95	921	1.64	27%
1987	3,399	0.94	1,000	1.78	29%
1988	3,510	0.97	1,131	2.02	32%
1989	3,430	0.95	1,024	1.83	30%
1990	3,319	0.92	1,051	1.87	32%
1991	3,298	0.91	NA	NA	NA
1992	3,422	0.95	NA	NA	NA
1993	3,564	0.99	1,373	2.45	39%
1994	3,563	0.99	1,578	2.81	44%
Average annual % change 1983–1994	-0.1%		9.9%		10.0%

NOTE: "U.S. shipments" refers to total shipments by manufacturers located within the U.S., including units to be exported. Cold-cathode fluorescent lamps are excluded from the U.S. shipment data; Christmas tree lights are excluded from U.S. shipments as well as imports.
NA = Not Available
SOURCE: U.S. Census Bureau current industrial reports MQ36B, various years
CONTACT: Jonathan Koomey, JGKoomey@lbl.gov. <http://enduse.lbl.gov>
FILENAME: shipmentsandimports.xls
DATE OF CREATION: February 1996
DATE OF LAST MODIFICATION: November 1997

Table 36.4 shows how to revise the table. I eliminated all but the essential gridlines and included indices relative to 1983 (which show how much shipments and imports grew from 1983 to 1994). I also added a column with imports normalized as the percentage of U.S. shipments, to help the reader understand how this important statistic changed over time. In addition, I included an additional row with the annual average percentage change from 1983 to 1994 because this summary statistic is helpful for understanding long-term trends in the data. Finally, I added information to the footnotes to identify the contact person, the file name, and the dates of creation and last modification for the file.

TABLE 36.5. A table that needs revision from another of our technical reports

Primary energy use in quads: 1973-1997

	1973	1986	1990	1995	1997
Buildings	24.1	26.9	29.4	32.1	33.7
Industry	31.5	26.6	32.1	34.5	32.6
Transportation	18.6	20.8	22.6	24.1	25.5
Total	74.2	74.3	84.1	90.7	91.8

SOURCE: Energy use estimates for 1973-95 come from EIA (1996a, Table 1.1, p. 39). Energy use estimates for 1997 come from EIA (1996c).

TABLE 36.6. Improved version of Table 36.5 for use in a summary report

Primary energy use in quads: 1973-1997

	1973	1986	1990	1995	1997	Average annual change 1973-1997 % per year
Buildings	24.1	26.9	29.4	32.1	33.7	1.4%
Industry	31.5	26.6	32.1	34.5	32.6	0.1%
Transportation	18.6	20.8	22.6	24.1	25.5	1.3%
Total	74.2	74.3	84.1	90.7	91.8	0.9%
Index (1973 = 1.00)	1.00	1.00	1.13	1.22	1.24	

SOURCE: Energy use estimates for 1973-95 come from EIA (1996a, Table 1.1, p. 39). Energy use estimates for 1997 come from EIA (1996c).
CONTACT: Jonathan Koomey, JGKoomey@lbl.gov. <http://enduse.lbl.gov>

Table 36.5 shows a third example, taken from another technical report that my group worked on.[83] In this case, the client insisted on a particular format for the table, but **Table 36.6** shows how I'd change the table if it were up to me (it's not bad, but it could be better). To improve it, I added a column containing the average annual percentage change from 1973 through 1997. I also added a row with an index relative to 1973, which (for example) shows at a glance that primary energy use in the U.S. grew 24% over this period. Finally, I added contact information, so readers with questions know whom to approach. The more detailed information with the file name and date of creation and last modification makes sense for a technical table but would be excessive for a summary table.

Table 36.7. improved version of Table 36.5 for use in a
detailed technical report

Primary energy use: 1973-1997

	1973	1986	1990	1995	1997
	Quadrillion Btus of primary energy				
Buildings	24.1	26.9	29.4	32.1	33.7
Industry	31.5	26.6	32.1	34.5	32.6
Transportation	18.6	20.8	22.6	24.1	25.5
Total	74.2	74.3	84.1	90.7	91.8
	Sectoral energy as a percent of total				
Buildings	32%	36%	35%	35%	37%
Industry	42%	36%	38%	38%	36%
Transportation	25%	28%	27%	27%	28%
Total	100%	100%	100%	100%	100%
	Index 1973 = 1.00				
Buildings	1.00	1.12	1.22	1.33	1.40
Industry	1.00	0.84	1.02	1.10	1.03
Transportation	1.00	1.12	1.22	1.30	1.37
Total	1.00	1.00	1.13	1.22	1.24
	Average annual percentage change since 1973				
Buildings	0%	0.8%	1.2%	1.3%	1.4%
Industry	0%	-1.3%	0.1%	0.4%	0.1%
Transportation	0%	0.9%	1.2%	1.2%	1.3%
Total	0%	0.0%	0.7%	0.9%	0.9%

SOURCE: Energy use estimates for 1973-95 come from EIA (1996a, Table 1.1, p. 39). Energy use estimates for 1997 come from EIA (1996c).
CONTACT: Jonathan Koomey, JGKoomey@lbl.gov. <http://enduse.lbl.gov>
FILENAME: primaryenergyuse.xls
DATE OF CREATION: January 1999
DATE OF LAST MODIFICATION: December 1999

Table 36.5 appeared in a summary chapter of a larger report, so the recommended revisions are slightly different than if the table were to appear in a more detailed technical chapter (**Table 36.7** shows a version that would be appropriate in that situation). The same information on primary energy use in the different sectors is shown in that table, supplemented by several indices. First, the table shows energy use by sector expressed as a fraction of the total. Next, it shows indices relative to 1973, making growth over the analysis period readily apparent. The table also includes average annual percentage rates of change since 1973 for the sectors and the total. Finally, the additional detail about the file name and dates of creation and last modification are appropriate here.

Although some aspects of creating good tables are matters of style and preference, the principles described above should remain foremost in your mind, because they are valid for almost every table. Readers should treasure your tables for their accuracy and completeness. Sloppy tables sap your credibility—avoid them at all costs.

There are two ways of constructing a software design [or a table]: one way is to make it so simple that there are obviously no deficiencies; the other is to make it so complicated that there are no obvious deficiencies. — **C. A. R. HOARE**

USE NUMBERS EFFECTIVELY IN ORAL PRESENTATIONS

Charles Osgood has a simple rule about using numbers in oral presentations: DON'T!

I disagree. You can use them sparingly, in carefully chosen settings. It is true, however, that most people can't absorb numbers and their implications without considerable hand-holding.

If you feel that numbers will enhance your argument, then summarize the few key numbers in graphic form. Only rarely use a table, and only then when it's a small one that contains the key information in large, readable type. If you have detailed data and calculations supporting your analysis, then give hard-copy handouts to your listeners. Summarize the conclusions from the results and focus on the conclusions and the story you are trying to tell, not on the numbers themselves.

Common doctrine is that data are dull. But as Edward Tufte points out, "if the statistics are boring, then you've got the wrong numbers" (or perhaps you are conveying them in a boring way). You should use numbers to make an important point. If your audience is not yet convinced, then *you* are responsible for presenting the relevant information in a way that is engaging to them. Data are dull only when chosen poorly and presented badly.

Make sure your numbers are correct! Nothing destroys your credibility faster than presenting numbers that are internally contradictory or obviously wrong. Once you lose credibility, it's nearly impossible to regain it during the course of a presentation. It won't matter how good the rest of your work is if you've presented bad data during the talk. So before you finalize your presentation, *dig into the numbers* and make sure they are both relevant and correct. Practice the talk before a friendly audience of colleagues, so they can catch any errors in advance.

Never display unreadable, densely packed tables or charts in presentations. I've seen numerous veteran professors do it, but it's insulting to the audience,

and it's guaranteed to make your listeners tune out. These speakers are not trying to be insulting, they just assume that others find raw numbers as compelling as they do. Don't fall into this trap. You're the expert, and you should pre-digest the information and put it in an understandable form. It's not just common courtesy, it's also the best way to ensure that your main points are most effectively conveyed.

Gerald Jones, in his book *How to Lie with Charts,* gives guidelines on the size and kind of text to use in presentations and describes the conventions for measuring font size and spacing. He quotes the Society of Motion Picture and Television Engineers, who recommend that the minimum height of small letters used in on-screen text should be 2.5 percent of the height of the screen. Typical presentation screens are 1.5 meters (5 feet) in height, so the minimum font size in such a screen should be about 4 centimeters (1.5 inches). These guidelines don't just apply to text presentations. Make sure the labels on your tables and graphs are just as readable as your bullet points.

Another way to judge type size of an overhead projector slide is to place it at your feet. If you have good eyesight and can't read it while standing, the type is too small.

A successful speaker always thinks carefully about the following questions:

- Who will be in the audience?
- What is important to them?
- What are the three key messages you want people to learn from your talk? (With rare exceptions, that's all they will ever take away.)

32

Although Ross Perot's use of charts and tables did not earn him electoral victory in 1992, he demonstrated that a speaker who addresses issues the audience cares about can be effective and entertaining even if he uses more graphs per minute than in any television show in history.

COMMENTING ON THE PRESENTATIONS OF OTHERS

If you participate in conferences, you will occasionally be called upon (or may feel obliged) to comment on results presented in the oral presentation of another analyst. The same rules apply, but there are some subtleties.

29 Use back-of-the-envelope calculations to assess their analysis during the talk, but take extra care that your calculations are correct before speaking because you will likely be doing these calculations under time pressure.

Be polite! Not everyone accepts public criticism gracefully, even when it's gentle criticism. It's often difficult to engage in reasoned debate in a public forum, so in some cases it will be more fruitful to talk in private with the speaker after the presentation.

CONCLUSIONS

Effective presentations tell a story. Use carefully chosen numbers to tell that story in a compelling and succinct way. Consider what the audience cares about as you prepare your remarks, and make sure your talk gives your listeners the information they need and the means to find out more.

When speakers use a lot of numbers, the audience most always slumbers. — **CHARLES OSGOOD**

USE THE INTERNET [84]

Anyone publishing analysis results in the 21st century had better take heed of the Internet. It is fast changing the way information is collected, processed, distributed, and used. Those who put the new medium to use will reap tremendous benefits; others ignore it at their peril.

FEATURES OF THE NET

In some ways, recent developments in electronic information distribution are a logical outgrowth and extension of the advent of mechanical type and printing presses five centuries ago. As James Burke points out in his book *Connections*, the trends reinforced by mechanical printing included rapid and accurate reproduction of texts, the specialization of knowledge and the responsibility of scholars for its accuracy, the first steps towards our current conception of intellectual property, and a sharp decline in the cost of reproducing and distributing information. The Internet accentuates each of these trends, but it also has its revolutionary aspects:

- *Near-zero marginal cost of reproduction and distribution:* Conventional printing vastly reduced the incremental costs of reproducing and distributing books, but the Internet reduces this cost to nearly zero. This shift from low marginal cost of reproduction to nearly zero cost is forcing another redefinition of our concept of intellectual property.

 Many companies are using the near-zero marginal cost of the Internet to reduce the cost of delivering customer support for their products. Frequently asked questions and product literature are now commonly made available electronically, which reduces the number of phone calls to customer support and therefore minimizes costs of fielding these inquiries.

It also allows the technical support staff to focus on difficult problems with fewer distractions. Shipping companies like Federal Express and UPS use the Internet for on-line package tracking, which vastly reduces their need for phone support.

- *Quicker publishing:* The barriers to publishing are now much lower because publishing materials on the net (and changing them) is relatively cheap, easy, and quick. The author Robin Williams summarizes Internet publishing as being "like a commercial print shop saying 'We'll print your recipe book this month for $30. Use as many colors and pages as you like. Change the recipes every day if you want, and we'll reprint it for you everyday, for free.'" This kind of flexibility does not exist in the world of conventional printing.

 On the other hand, lower barriers to entry in Internet publishing place a higher responsibility on each publisher. The Internet is a more freewheeling place than conventional media sources, and it lacks the institutional credibility checks that are prevalent in newspapers, radio, television, and conventional publishing. It is therefore incumbent on each web publisher to establish and maintain its credibility in any way it can. Testimonials from established organizations with credibility can help with this effort, as can clear and exhaustive documentation of sources and methodology.

- *Quicker review of technical material:* For analysis projects, making spreadsheets and draft reports available to readers on the Internet is a real boon. If readers have the spreadsheets containing the complete calculations, they can delve into the analysis much more easily and quickly than if they merely received a copy of the report. Full electronic access by end users speeds up the rate at which analysis can be reviewed, which reduces the time needed to create a final analysis.

- *Easier ordering and distribution:* Selling hard copy books and reports is easier and cheaper than in the days before the Internet. For certain kinds of books, at least, it is no longer necessary to have them stocked in local bookstores: people can find and order them on line, and receive them in a few days. This feature of the Internet eliminates a major hurdle for authors and small publishers.

- *Direct feedback from suppliers to consumers:* For manufactured products, many companies give an estimate of shipping time for a particular product on their web sites. The most sophisticated companies have a direct link between their computerized inventory management system and the web

store that contains the product, which always reflects the current product supply. Never before have consumers known details about product supply and timing, and this capability gives them more flexibility in their purchasing decisions.

- *Direct feedback from consumers to suppliers:* Some companies are experimenting with using the Internet to allow their customers to supply specific information about company performance. The on-line auctioneer Ebay <http://www.ebay.com>, for example, has what it calls a "Feedback Forum," where buyers and sellers can comment on each other and on Ebay itself.

- *Indirect feedback from consumers to suppliers:* The Internet offers more and cheaper information about customers than ever before. Customers can specify the exact products they want, and this information is automatically collected in a database that can be "mined" by the company to learn about its customers. The company can also use information about who accesses its web site to create more targeted marketing campaigns. In addition, the company can track how people are using the search engine on the site to find out what kinds of questions people are trying to answer when they visit the site. This information can help the site developer improve the site but may also lead to ideas for new products to meet user needs.

- *Collaboration among users:* Well-designed collaborative web sites can allow customers and users to comment on the information distributed by the web publisher and to have those comments visible to others who visit the site. Such sites can also help users help each other by setting up discussion forums on topics related to the company's products. Interested users who have a specific question can visit such forums and have their questions answered by other interested customers, at little cost to the company itself beyond initially setting up the site. These user forums can also be a fertile source of ideas for the company's product designers on improving existing products and creating new ones.

- *Access to information 24 hours per day:* Many information requests no longer need be made during business hours. Searching on the net can answer many questions at any time of the day or night. This flexibility reduces costs for all companies and saves users time, which benefits everyone. It is particularly beneficial when the company and the customer are separated by many time zones.

- *Universal searching:* The Internet allows access to vast storehouses of knowledge from all over the world and gives automated ways (search engines) to explore it. Never before has it been feasible to search such a vast network containing so much information. Finding essential data is often as much art as science, but experience with and study of search engine algorithms can help users find information quickly.

- *Easier and more widespread public access to technical information:* Technical information is often buried in impenetrable and obscure reports. Interactive links between the Internet and relational database management systems help those who possess detailed technical knowledge to make it useful to a wider audience.

Lawrence Berkeley National Laboratory (LBNL), for example, has for years been the preeminent center on energy use in homes, but much of the information LBNL generated never reached the general public until the advent of the World Wide Web. LBNL's Home Energy Saver web site <http://HomeEnergySaver.lbl.gov> was the first Internet-based home energy analysis tool; it embodies the technical expertise of dozens of people throughout LBNL. Users of this site have confidence that the tool accurately characterizes energy use in their home because of the expertise and credibility of those who created it.

The Environmental Defense Fund's Scorecard web site <http://www.scorecard.org> allows everyone to determine the sources of pollution in their own communities merely by entering their zip code. It identifies the companies that produce the pollution and even lets users send a fax to the polluters complaining about it. The technical content of the site was created by people whose expertise in environmental issues is well known. They relied on publicly available data and compiled it in a form that any non-technical person could use to get immediate and understandable results. The information was available before the Scorecard team pulled it together, but no one had ever made it accessible in such an easy-to-use form.

The increasing availability of information about pollution, energy costs, political decisions, campaign contributions, and the spending habits of all types of institutions will have a larger and larger influence on our society during the next decade. Once the data are compiled, they can be widely distributed at near-zero marginal cost. People need no longer take at face value the assurances of community leaders about top-

ics of current interest. They can see and decide for themselves how to respond.

These features of the Internet present opportunities for analysts who are willing to seize them. The next section presents some specific advice for those who want to use the Internet to distribute data and analysis results with greatest effectiveness.

PUBLISHING ANALYSIS IN THE INTERNET AGE

As of Spring 2000, there are two essential books on web publishing for analysts to read. The first is *The Non-Designer's Web Book,* by Robin Williams and John Tollett <http://www.peachpit.com/>, which contains reams of practical advice about making the Internet work for you. It gives information about color, file formats, typefaces, searching strategies, and graphics, and describes how Williams' design principles (from *The Non-Designer's Design Book*) should be applied to web sites. The second is *Philip and Alex's Guide to Web Publishing,* by Philip Greenspun <http://philip.greenspun.com/panda/>, which is a relatively advanced guide to the mechanics of web publishing, focusing heavily on techniques to enable collaboration between people and institutions. It is exemplary for its no-nonsense, sensible approach to dealing with complex issues, and it is also amusing. I strongly recommend you read both books cover to cover (even if you skip the really technical parts of Greenspun's book, it's worth reading).

In the section that follows, I assume familiarity with the basics of web publishing, so I can focus on key lessons of most relevance to analysts. For an introduction to the basics of the Internet, refer to Parts 1 and 2 in the Williams and Tollett book as well as the following page on Yahoo.com, which presents resources for beginning web designers: <http://dir.yahoo.com/Computers_ and_Internet/Internet/World_Wide_Web/Beginner_s_Guides>. Here's my specific advice:

Know your audience: This is the first rule of all presentations, whether written or oral, and it applies just as well to web sites. Decide who will use your site, and make sure they can find what they want easily and quickly.

32

Focus on content, not graphics and fancy gimmicks: Many web designers are obsessed with graphics and the latest programming tricks. Including lots of graphics ensures that people using slower connections (the majority of users) will spend a lot of time waiting. Relying on the latest web programming tricks ensures that people with older browsers won't be able to use your site (it also ensures that you will have to field technical support questions from users who can't get your web page to work on their computer).

Web sites should convey information, not display technical virtuosity. Determine the simplest approach to get your message out, and stick to it. Value your user's time and design your site so that anyone can quickly and efficiently learn what you have to teach. One way to maintain this focus on the user's needs is to assign responsibility for content and usability to someone different from the web designer.

Do usability testing: Web sites are more like software than printed material. Even if you think you know your audience, you need to test the site in action. Greenspun recommends that you observe a few selected people using your site to accomplish specific tasks, which can help you "anticipate user questions and make sure your site answers them." Follow the guidance on "Involving users in the design process" found in the *Macintosh Human Interface Guidelines* (which is a time-tested source on how to make computers work well for humans). User testing is the best way to ensure that documentation is sufficient and that your user interface design is smooth as silk.

Document like crazy: Documentation is just as important for web pages as for books and hard-copy reports. Each page should have a "Date last modified" at the bottom as well as links that send email to the person who made it and to others responsible for the technical material described on the site. Don't forget that web sites need to be backed up just like all other electronic information, both to safeguard the current version and to document past versions.

Use PDF: Portable Document Format (PDF) is a universally available and widely used format for distributing documents electronically. Readers can print PDF files that are exactly the same as the original report, which makes distributing reports electronically practically the same as mailing the reports (par-

ticularly as high-speed duplexing laser printers become more common). More sophisticated PDF files contain internal hypertext links, so you could jump from a listing in a table of contents directly to the relevant place in the document. PDF files can and should be made available for download from your web site.

Adobe Acrobat, the program used to create such files, is available for Windows and Macintosh computer platforms <http://www.adobe.com/products/acrobat/>. The Acrobat reader is free and it allows users of almost every type of computer, including Windows, Macintosh, Linux, and most forms of UNIX, to read and print PDF files <http://www.adobe.com/products/acrobat/readstep.html>.

Schemes for selling electronic content are still in nascent form, so there is a tension in the Internet world between free distribution (wide availability) and being rewarded for your efforts. To date, no proposed methods for addressing these issues have caught on. One clever approach was championed by the now defunct company called Softlock.com. Its system allowed anyone to view and email a sample of a given PDF document. To see the full document, the reader had to go to the Softlock.com web site and pay for access. Once the fee was paid, the full version of the PDF file was unlocked for use on the user's computer. If the file was emailed or copied to another computer, it reverted back to the sample so that the recipient of the file could evaluate it and decide whether it was worth buying. The creator of the file could specify the level of access that purchase of the file grants (read only, read and print only, read/print/copy text, etc.). Unfortunately, the technology did not achieve market success.

Assign one web page per publication: Assign every new publication a unique web page, and write the URL on the title page of the document. This web page should include related data, analysis spreadsheets, related links to other people's sites, errata discovered after publication, and (in many cases) downloadable PDF files of the report itself. Owners of the hard-copy report know that they can always go to that page to get further info. They can also direct other interested people to that web page to avoid having to make a copy themselves. This approach requires a file structure that does not change after the site is designed, so make sure you've settled on a sensible structure before widely publicizing your site (this is good advice in any case).

Make data and calculations available: For web pages from which users will access data files, post downloadable spreadsheet files with plenty of description of the files on the web site, including file format, name, version number, date last modified, and size. The link to download a particular file might then look like this: *downloadme.xls* (150k), version 2, in Microsoft Excel 97/98 format (last modified 31 October 1999). This information can help a user decide whether to download the file. That user may have a slow network connection, for example, and may not want to download files greater than 1 Megabyte, or may not have Microsoft Excel. By giving users the file size and format, they can make the decision that is most appropriate for each situation. Be sure to document your calculations and data sources in the file itself so that you avoid readers badgering you with questions that should have been answered by your footnotes in the first place.

From the perspective of making data available for calculations, it is almost always better to have a downloadable file than an HTML page that you must save as text (of course the HTML page is easy and quick to read). If you have the time, it's best to make both available to your on-line readers, to serve both the need for quick access to some of the data and the need for easy spreadsheet access to all the data. Cleaning the stateline.org data would have been a lot quicker with a downloadable file.

Use repetition to good advantage: Develop a uniform format for the web site's pages. This approach relies on the same principle that makes Tufte's "small multiples" so effective in conveying information: once readers understand the format, they need not re-examine it each time they access the information anew. On my group's site at LBNL <http://enduse.lbl.gov>, we use the same navigation toolbar at the top of every page. It looks like this:

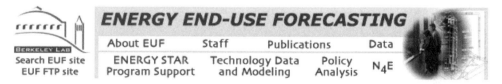

We also have a text toolbar at the bottom of every page that replicates many of the functions of the graphical toolbar, for people with slow connections who browse the web with graphics turned off.

The text toolbar looks like this:

About EUF | Staff | Publications | Project Areas | Data | End-Use FTP Site | Search

Wherever readers find themselves on the site, they always have these toolbars available and can access the information they need quickly and effectively.

The same principle applies to the content of the pages themselves. On my group's web site, we have unique pages for each distinct project (for example, see <http://enduse.lbl.gov/Projects/InfoTech.html>). Each project page has consistent headings:

Project Description, which gives a few paragraphs describing the work and an occasional figure to illustrate important results;

Project Staff, which is a list of those who worked on the project with a direct link to each person's biography page and contact info;

Key Data, which includes downloadable files that contain inputs and results for the analysis;

Publications, with full references to the reports as well as links so that the complete PDF files can be downloaded by the reader;

Other Resources, which includes links to external sites related to the project;

Sponsors, which lists the people and institutions responsible for funding the project.

Readers know that the project pages all have these headings; this familiarity makes it easier to find relevant information quickly.

Pack your site with carefully chosen text: Search engines look for text, so the more text on your site, the better (this is another reason to focus on textual content and not on graphics). Of course, your pages should attract readers and help them find the information they need, so follow Williams' advice about proximity, alignment, and contrast in this effort. Don't overwhelm readers, but do put as much content on your site as makes sense. Greenspun suggests that you consider purchasing the rights to text that is related to the purpose of your site and putting it up. This approach is cheaper and more likely to attract visitors than purchasing on-line advertisements on other sites (anyone using a search engine to find information on a topic related to that of your site will have many links to your site pop up from their search query).

If you want to attract search engines to a page that is predominantly graphical, HTML allows programmers to include "meta-tags" that contain key words summarizing the purpose of the page and site. Another trick is to put text on the bottom of that page that is white on white background. It won't appear to those viewing the page with a browser, but the search engine will use that text to index the page.

Enable collaboration: One of the most intriguing and powerful aspects of Internet publishing is the potential for on-line collaboration. Web sites that encourage readers to add comments, ask questions, and answer queries become a treasure trove of information, not just for the users of the site, but for the publisher as well.

Another kind of collaboration is enabled by the use of "Project sites" <http://www.secretsites.com/>, which contain the project's prospective schedule, an email trail, the historical chronology of project developments, and the relevant electronic documents. Creating such a password-protected site makes the key information available to all participants in the project and allows them to work together more effectively.

Interact with users: Greenspun says "User feedback might be annoying at times, but if you want to be insulated from your readers, why publish on the Web at all?" The Internet allows closer connection between customers and businesses than previously possible. Use those connections to your advantage, and learn about your customers to serve them better. Collaborative services can help in this learning process, as can analysis of server logs and direct email interaction. Give users the ability to comment on your site and your products, and you are certain to learn things you could not have learned on your own.

Set up an internal search engine and site index: A search engine devoted to your site will save your readers time and frustration. Greenspun suggests that you program it to send you email if a search returns no items or is repeated several times: "As soon as two users ask the same question, you should beef up your site to answer it unattended." Thus, the search engine can help you improve your site, assist your users, and teach you about their needs all at the same time. A site index, which lists keywords with links to the related part of the site, will also help readers find the information they need.

Get listed in external search engines: Make sure you're registered with the major search engines and directories because these services are likely to be your most important source of web traffic (there are companies that do this registration for you, for free or for a nominal fee). Such registration insures that people searching for the information you have will be able to find you quickly. Some of the most important search engines (in alphabetical order) are Altavista, Excite, Google, HotBot, Infoseek, Lycos, Webcrawler, and Yahoo. For the latest information on this process, check out the listings at <http://dir.yahoo.com/Computers_and_Internet/Internet/World_Wide_Web/Site_Announcement_and_Promotion/> or the advice available at <http://www.searchenginewatch.com>. Williams and Tollett point out that some search engines specialize in certain types of technical information. If there's one for your specialty, then make sure you register with that one as well.

Use printed material when appropriate: One downside to the Internet (as of Autumn 2000), is that reading text on a computer screen is still not as convenient or as pleasant as reading a book. Printed documents can be read anywhere and have a clarity and permanence that electronic media cannot yet match. Until the technology for electronic books becomes better developed, paper books and reports will remain important (though *distribution* of some of these documents will surely be replaced by PDF files that can be electronically distributed).

Link to other sites: Reports that only reference the author's own work rightfully encourage skepticism, and web sites that don't refer to other kindred sites deserve similar treatment. Add value to your site by helping readers find other sites that provide complementary information and data. Access to comprehensive lists of related links is one of the reasons web surfers return to a site again and again.

Bring in the big guns: When you have a lot of data that you want to make available, call in the experts on database-backed web sites. Greenspun describes the trials, tribulations, and triumphs of the people who make these complex sites happen. The Home Energy Saver and Scorecard web sites are two good examples of how this kind of programming can make complex data and analysis results accessible to a wide audience. These sites can be quite expensive to develop and maintain but can be indispensable for certain applications.

CONCLUSIONS

Seize the opportunities the Internet presents. After defining your audience, create content that matters to them and make it available on a well-designed and well-tested site. Avoid fancy graphics because they consume bandwidth and contribute little to getting your message across. Make sure your site is accessible to people using modems, older browsers, and all major operating systems. Set up comment pages and discussion forums to help your readers teach you and each other. Finally, remember that the Internet is just another tool and that it's not *always* the right tool for the job. Non-electronic publishing still has its place.

When I am working on a problem I never think about beauty. I only think about how to solve the problem. But when I have finished, if the solution is not beautiful, I know it is wrong.

— **BUCKMINSTER FULLER**

CONCLUSION

CREATING THE FUTURE

Like the butterfly's wing beat that ultimately leads to a hurricane, seemingly trivial actions can have consequences that reverberate through generations. With every action, with every day we live, we create the future. Of course, forces beyond human control also have influence, but it is how our choices relate to these external events that determine the outcome.

Mastering the art of problem solving is one important way to ensure that the future you create is a hopeful one. Analysis helps you make choices that are consonant with your personal values, consistent with your ideology, and reflective of the realities imposed by external forces. Whether you are a businessperson, scientist, athlete, or accountant, analyzing your choices in a systematic way will lead to a better life.

Of primary importance from this book are the following lessons:

- *Don't be intimidated by anyone:* The key issues surrounding all but the most technical topics can be understood by any intelligent and sufficiently diligent person. `2`

- *Be a critical thinker:* Dissect your own arguments and those of others. Identify the premises and the main conclusion of each argument. Make sure the premises are acceptable, relevant, and adequate to support the conclusions. Finally, search for missing arguments and for counter arguments. `11`

- *Don't confuse what's countable with what really counts:* Many of the most important things in life can't be quantified, so don't focus just on the numbers—they aren't everything. `12`

- *Get organized:* Clean up your office. Establish a filing system that allows you to find what you need when you need it. Buy the key books in your field, and make them accessible. Above all, remember that organization often beats brilliance! `6` `7`

- *Question authority:* Maintain a healthy skepticism of other people's analysis, and ask questions until you understand their points. Clearly dis-

19 tinguish facts from value judgments, and make sure that others do so too.

16 Don't believe everything you read.

27 • *Dig into the numbers:* Look for internal contradictions, large differences, obvious trends, and cognitive dissonance. Compare results to independent sources. Take ratios of numbers, and make sure they relate in a predictable way.

- *Focus on the essential:* In creating your analysis, focus on designing consistent comparisons, developing credible scenarios, and using simple and understandable models to illustrate the key issues. Avoid complex computer tools unless absolutely necessary. **25 26** **28**

- *Document, document, document:* Documentation creates a trail for you and others to follow after the analysis is done. It is essential to any intellectual work but is oft neglected. By making it a priority, you ensure that your intellectual contribution will remain important and useful for years to come. **33**

- *Use the Internet:* Rely on the Internet to conduct your research and make your publications widely available. The old ways of publishing and disseminating data are rapidly being swept away by the new electronic media, so use these new techniques to your advantage. **38**

- *Remember that others don't care as much about your work as you do:* It's your responsibility to know your audience, address their concerns, and explain your work in an understandable way. Make your figures clear and compelling, your tables well documented, and your reports so valuable that your colleagues will treasure them for years. **32** **35** **33 36**

- *Synthesis follows analysis:* Combining analysis results from different fields in creative ways often leads to new insights. Such synthesis is one powerful way to increase your understanding of the world around you. Don't underestimate its importance.

Finally, the integrity of your calculations and the clarity of their presentation are both critical to your analytical success, but even more important is your focus on the issues that really matter. Too many people expend their effort on topics of only marginal importance. Your time is your life, so make it count! **10**

I leave you with the environmentalist Edward Abbey's excellent advice for those who are trying to change the world: "Do not burn yourself out. Be as I am—a reluctant enthusiast, a part-time crusader, a half-hearted fanatic. Save the other half of yourselves and your lives for pleasure."

You can't accomplish your goals unless you yourself are whole. Care for yourself and those around you. Take the time to do those things that replenish your energy and fill your heart with joy. Your children and our posterity will thank you for it.

The future is where we will spend the rest of our lives.

— ANONYMOUS

FURTHER READING

NUMBERS TO KNOWLEDGE "TOP TEN" LIST

Choosing the top ten from the "Further Reading" section is a subjective exercise, but I'm sure you won't be disappointed with any of these:

1. To learn about searching on the Internet, see <http://www.searchenginewatch.com> and <http://dir.yahoo.com/Computers_and_Internet/Internet/World_Wide_Web/Searching_the_Web>

2. Booth, Wayne C., Gregory G. Colomb and Joseph M. Williams. 1995. *The Craft of Research*. Chicago, IL: The University of Chicago Press. This superb book presents a framework for dissecting arguments that can be applied in most situations.

3. Cialdini, Robert B. 1993. *Influence: The Psychology of Persuasion*. New York, NY: Quill–William Morrow. Cialdini's book shows how "compliance professionals" use their knowledge of the human psyche to influence people's decisions.

4. Huff, Darrell. 1993. *How to Lie with Statistics*. New York, NY: W. W. Norton & Co., Inc. If this book doesn't make you a skeptical observer of numbers and statistics, nothing will. First published in the 1950s, it's still fresh and useful today.

5. Ueland, Brenda. 1987. *If You Want to Write: A Book About Art, Independence, and Spirit*. St. Paul, Minnesota: Graywolf Press. There is no better guide to sparking your creative process.

6. Hughes, William. 1997. *Critical Thinking: An Introduction to the Basic Skills*. 2d Peterborough, Ontario: Broadview Press. This clearly written and accessible book is a marvelous introduction to logical argument.

7. Norman, Donald A. 1990. *The Design of Everyday Things*. New York, NY: Doubleday/Currency. This fascinating book explores how humans interact with technology, and gives guidance to designers on how to make their products a joy to use.

8. Schwartz, Peter. 1996. *The Art of the Long View: Planning for the Future in an Uncertain World*. New York, NY: Doubleday. Lively and realistic scenarios are crucial tools for decision makers, and Schwartz describes how to create them in this classic book.

9. Tufte, Edward R. *Envisioning Information*. (1990). *The Visual Display of Quantitative Information*. (1995). *Visual Explanations: Images and Quantities, Evidence and Narrative*. (1997). Cheshire, CT: Graphics Press. OK, so I'm cheating a bit counting these three as one, but they're so good you'll want them all. <http://www.edwardtufte.com>

10. U.S. Bureau of the Census. 2006. *The Statistical Abstract of the United States 2006*. 125th edition Washington D.C.: U.S. Government Printing Office. One of the most important data sources for any U.S. researcher. The plot's a bit thin, but the content's superb.

INTRODUCTION • THE INFORMATION EXPLOSION

Burke, James. 1978. *Connections*. Boston: Little, Brown and Company. An outstanding book that discusses the history of technology in a brilliant and accessible way. Among other things, Burke explores the key role of the printing press in fostering later technological and scientific developments.

de Sola Poole, Ithiel. 1983. *Technologies of Freedom*. Cambridge, MA: The Belknap Press of Harvard University Press. This book made a first attempt to quantify the accelerating pace of information generation in modern society.

Scientific American. 1997. *Scientific American, Special Issue: The Solid-State Century–The Past, Present, and Future of the Transistor*. vol. 8, no.1. A useful compilation of recent information on developments in electronics and information technology.

Shenk, David. 1997. *Data Smog: Surviving the Information Glut*. New York, NY: HarperCollins Publishers, Inc. This book documents the symptoms of information overload. It's worth a read just for the anecdotes, which are delightful.

PART I • THINGS TO KNOW

1 • Beginner's Mind

Hyams, Joe. 1982. *Zen in the Martial Arts*. New York, NY: Bantam Books. The story of Joe Hyams' explorations into different martial arts is one of the inspirations for the book you are now reading. It's broken up into short sections, each with an important lesson he learned over the years. A quick read, and a great one.

2 • Don't Be Intimidated

Ringer, Robert J. 1974. *Winning Through Intimidation*. New York: Funk & Wagnalls. This book is one of my favorites. It's amusing, entertaining, interesting, and timeless. It's also a quick read, and is available in most big bookstores (it still sells well).

Ueland, Brenda. 1987. *If You Want to Write: A Book About Art, Independence, and Spirit*. St. Paul, Minnesota: Graywolf Press. One of the classic books on creativity. Ueland writes about writing, but her advice applies equally well to most forms of human creation. It's beautifully written and easy to read.

3 • Information, Intention, and Action

Hammond, John S., Ralph L. Keeney, and Howard Raiffa. 1999. *Smart Choices: A Practical Guide to Making Better Decisions*. Boston, MA: Harvard Business School Press. A well-written and methodical guide to improving your decision making processes.

Hess, Peter, and Julie Siciliano. 1996. *Management: Responsibility for Performance*. New York, NY: Mc Graw-Hill, Inc. This book does a fine job of describing how to set goals, make decisions, and implement decisions in the business world. It also explicitly discusses the role of intuition, experience, and values in decision making.

Kawasaki, Guy. 1995. *How to Drive Your Competition Crazy: Creating Disruption for Fun and Profit*. New York, NY: Hyperion. Kawasaki harkens back to his time as chief evangelist for

the upstart Apple Computer in the late 1970s and early 1980s, and gives some lessons for marketing products in the face of competition from huge and well-funded rivals. The book is loaded with great anecdotes.

Laderman, Jeffrey M. 1998. "Wall Street's Spin Game: Stock Analysts Often Have a Hidden Agenda." In *Business Week*. October 5. pp. 148-156.

Morgan, M. Granger, and Max Henrion. 1992. *Uncertainty: A Guide to Dealing with Uncertainty in Quantitative Risk and Policy Analysis*. New York, NY: Cambridge University Press. A "must read" book for anyone trying to understand how uncertainty affects decision making.

Norman, Donald A. 1990. *The Design of Everyday Things*. New York, NY: Doubleday/ Currency. This fascinating book explores how humans interact with technology, and gives guidance to designers on how to make their products a joy to use.

Schwartz, Peter. 1996. *The Art of the Long View: Planning for the Future in an Uncertain World*. New York, NY: Doubleday. Lively and realistic scenarios are crucial tools for decision makers, and Schwartz describes how to create them in this classic book.

4 • Peer Review and Scientific Discovery

Edwards, Paul, ed. 1972. *The Encyclopedia of Philosophy*. New York, NY: Macmillan Publishing Co., Inc. & The Free Press. Volumes 1 through 8. This encyclopedia is the standard reference source for anyone investigating philosophical concepts. It contains articles by leading scholars on everything from logic to epistemology, energy to truth, liberty to idealism, ethics to virtue, and almost anything else you can think of.

Hughes, William. 1997. *Critical Thinking: An Introduction to the Basic Skills*. 2d Peterborough, Ontario: Broadview Press. This clearly written and accessible book is a wonderful introduction to logical argument. Hughes uses myriad examples to illustrate the logical fallacies that bedevil most arguments. He also clearly describes the difference between inductive and deductive logic.

ISI. 1996. *Journal Citation Reports: A Bibliometric Analysis of Science Journals in the ISI Database (Printed Guide to the JCR on Microfiche)*. Philadelphia, PA: Institute for Scientific Information. This report should be available at most large public libraries.

Kuhn, Thomas S. 1996. *The Structure of Scientific Revolutions*. 3rd ed. Chicago: University of Chicago Press. This book redefined contemporary views of how science progresses, and is a landmark in the history of science.

Pirsig, Robert M. 1981. *Zen and the Art of Motorcycle Maintenance*. NY: Bantam Books. This book is a popular excursion into the relationship between science and art. While not always philosophically rigorous, the book is still entertaining, thoughtful, and provocative.

Sokal, Alan D. 1997. "A Plea for Reason, Evidence, and Logic." *New Politics*. vol. 6, no. 2. p. 126. This short transcript of a talk by Sokal is a brilliantly clear and concise exposition of the scientific worldview, in the context of his now famous parody of "post-modernist" relativism. Sokal created a controversy in 1996 by submitting an article that he knew was total nonsense to a journal called *Social Text*, and the journal published it. Sokal suspected correctly that by liberally sprinkling the article with phrases that flattered the editors' predispositions, he would make publication likely. He then exposed the hoax. For the complete set of articles and related debates, see <http://www.physics.nyu.edu/faculty/sokal>.

PART II • BE PREPARED

5 • Explore Your Ideology

Flew, Anton, ed. 1984. *Dictionary of Philosophy*. London: Pan Books, Ltd.

6 • Get Organized

Covey, Stephen R. 1990. *The 7 Habits of Highly Effective People*. NY, NY: Simon & Schuster. Covey got me to think systematically about how I organize my time and my work.

Culp, Stephanie. 1990. *Conquering the Paper Pile-up: How to Sort, Organize, File, and Store Every Piece of Paper in Your Home and Office*. Cincinnati, Ohio: Writer's Digest Books. Practical advice for getting organized.

Lively, Lynn. 1996. *Managing Information Overload*. The WorkSmart Series. New York, NY: AMACOM/American Management Association. This book is brief but important. Its advice on controlling information flows is especially pertinent in the information age.

Morgenstern, Julie. 1998. *Organizing from the Inside Out: The Foolproof System for Organizing Your Home, Your Office and Your Life*. New York, NY: Henry Holt and Company. This is the best book on organizing that I've found. It's a top seller.

Pollar, Odette. 1996. *365 Ways to Simplify Your Work Life*. Chicago, IL: Dearborn Trade Publisher (800/829-7934). More practical advice to increase your effectiveness and reduce stress.

7 • Establish a Filing System

Culp, Stephanie. 1990. *Conquering the Paper Pile-up: How to Sort, Organize, File, and Store Every Piece of Paper in Your Home and Office*. Cincinnati, Ohio: Writer's Digest Books. A helpful book focusing specifically on dealing with paper clutter.

Lively, Lynn. 1996. *Managing Information Overload*. The WorkSmart Series. New York, NY: AMACOM/American Management Association. Advice on remaining organized in the face of a flood of information.

Morgenstern, Julie. 1998. *Organizing from the Inside Out: The Foolproof System for Organizing Your Home, Your Office and Your Life*. New York, NY: Henry Holt and Company. Has a fine chapter on "traditional offices and filing systems."

Schwartz, Peter. 1996. *The Art of the Long View: Planning for the Future in an Uncertain World*. New York, NY: Doubleday. See the section on "Information hunting and gathering".

8 • Build a Toolbox

Consortium for Mathematics and its Applications (COMAP). 1997. *For All Practical Purposes*. New York, NY: W. H. Freeman and Co. Project Director: Solomon Garfunkel. A truly outstanding textbook on using mathematics in the real world.

Gonick, Larry, and Woollcott Smith. 1993. *The Cartoon Guide to Statistics*. New York, NY: Harper Perennial, a division of Harper Collins Publishers. This book is a fun way to learn basic statistical concepts.

Harte, John. 1985. *Consider a Spherical Cow: A Course in Environmental Problem Solving*. Los Altos, CA: William Kaufmann, Inc. This beautiful book contains tools relevant to solving problems in environmental science. The sequel to it (*Consider a Cylindrical Cow: Further Adventures in Environmental Problem Solving*) was published in 2001 by University Science Books, in Sausalito, California.

Huff, Darrell. 1993. *How to Lie with Statistics*. New York, NY: W. W. Norton & Co., Inc. Reread this classic book every few years, to keep your toolbox fresh.

Huff, Darrell. 1996. *The Complete How to Figure It*. New York, NY: W. W. Norton & Co., Inc. An excellent sourcebook, with hundreds of examples of how to solve different problems in the real world.

Jones, Gerald E. 1995. *How to Lie with Charts*. San Francisco, CA: Sybex. A fine source for ideas on graphical and tabular presentation.

Kernighan, Brian W., and Rob Pike. 1999. *The Practice of Programming*. Reading, MA: Addison-Wesley. This marvelous book on the art of computer programming is somewhat technical, but is exceptional for its advice on programming style and documentation. It also offers many excellent suggestions for debugging and testing calculations.

Swartz, Clifford E. 1973. *Used Math: for the First Two Years of College Science*. Englewood Cliffs, New Jersey: Prentice-Hall, Inc. Summarizes mathematics that can be useful for more advanced analysts.

9 • Put Facts at Your Fingertips

Annual Reviews. 4139 El Camino Way, PO Box 10139. Palo Alto, CA 94303-0139. 650/493-4400. <http://www.annualreviews.org>

Berkman, Robert I. 1990. *Find It Fast: How to Uncover Expert Information In Any Subject*. New York, NY: Harper & Rowe, Publishers. A helpful guide to digging out data from the myriad disparate sources out in the world.

Blocksma, Mary. 1989. *Reading the Numbers: A Survival Guide to the Measurements, Numbers, and Sizes Encountered in Everyday Life*. New York, NY: Penguin USA. This book is somewhat hard to find, but it explains how to interpret and use a variety of data that we encounter in everyday life.

Brown, Lester R., Michael Renner, and Brian Halweil. 1999. *Vital Signs 1999: The Environmental Trends That Are Shaping Our Future*. New York, NY: W.W. Norton and Co. Tons of good time-series data on environmental trends. Worldwatch also sells these data in electronic form in both Windows and Macintosh formats. <http://www.worldwatch.org>

Edwards, Paul, ed. 1972. *The Encyclopedia of Philosophy*. New York, NY: Macmillan Publishing Co., Inc. & The Free Press. vols. 1 through 8. The best place to go when you need some background on an intellectual concept.

Famighetti, Robert, ed. 1999. *The World Almanac and Book of Facts 2000*. Mahwah, New Jersey: World Almanac Books. A "must have" reference. It comes out every year, and is chock full of useful information.

Lide, David R., ed. 1997. *CRC Handbook of Chemistry and Physics*. 78th ed. Boca Raton, Florida: CRC Press, Inc. The best source for physical constants and conversions. CRC also has many other technical and scientific reference works.

Simpson, John and Edmund Weiner, eds. 1989. *The Oxford English Dictionary*. Oxford, UK: Oxford University Press. 2d edition. This reference work takes up 20 volumes and 22,000 pages! This version costs $3,000, but you can get a complete version with *very* small type for about $350 (it comes with its own powerful magnifying glass). A better choice is the CD-ROM version, available for both Windows and Macintosh for about $400.

Sutcliffe, Andrea. 1996. *Numbers: How many, How Long, How Far, How Much*. New York, NY: Stonesong Press, Inc. Another good sourcebook for all sorts of helpful data.

U.S. Bureau of the Census. 1999. *The Statistical Abstract of the United States 1999*. 119th edi-

tion Washington D.C.: U.S. Government Printing Office. One of the most important data sources for any U.S. researcher.

WRI (World Resources Institute). 1998. *World Resources 1998-99*. Oxford, UK: Oxford University Press. Every two years, WRI compiles key environmental data from all over the world. This edition focuses on indicators of environmental change and human health, but it also has the core data on energy and environment that have been the hallmark of this publication for many years. There is also a companion CD-ROM that contains a huge array of related data. <http://www.wri.org>

10 • Value Your Time

Covey, Stephen R. 1990. *The 7 Habits of Highly Effective People*. NY, NY: Simon & Schuster. Covey is a highly successful "personal growth" guru who doles out advice for a living. His book helps you explore how your life patterns thwart or facilitate success.

DeMarco, Tom, and Timothy Lister. 1987. *Peopleware: Productive Projects and Teams*. New York, NY: Dorset House Publishing Company. A superb book that uses real-world experience to explode dozens of management myths. If you manage creative people, you owe it to them and to yourself to read this book.

Dominguez, Joe and Vicki Robin. 1993. *Your Money or Your Life: Transforming Your Relationship with Money and Achieving Financial Independence*. New York, NY: Penguin Books. Includes a clear and persuasive statement of how money, time, and your life are intimately connected.

Hyams, Joe. 1982. *Zen in the Martial Arts*. New York, NY: Bantam Books. The chapter titled "Do not disturb" describes how Hyams and his martial arts colleagues take command of their time.

Thich Nhat Hanh. 1991. *Peace is Every Step: The Path of Mindfulness in Everyday Life*. New York, NY: Bantam Books. This book advocates awareness in every waking moment of every day, so you never forget how amazing the world is. Sadly, most of us forget this all the time.

Thoreau, Henry David. 1971. *Thoreau: Walden and Other Writings*. New York, NY: Bantam Books, Inc. Joseph Wood Krutch, ed. Contains Thoreau's musings during his stay at Walden Pond.

PART III • ASSESS THEIR ANALYSIS

General

Huff, Darrell. 1993. *How to Lie with Statistics*. New York, NY: W. W. Norton & Co., Inc. There's no better exposé of commonly used tricks that mislead and misrepresent than this book.

Hughes, William. 1997. *Critical Thinking: An Introduction to the Basic Skills*. 2d Peterborough, Ontario: Broadview Press. Hughes presents examples and exercises to explore and categorize the logical fallacies that befall most arguments.

Lieber, Hugh Gray, and Lillian R. Lieber. 1944. *The Education of T. C. MITS*. New York: W. W. Norton & Company, Inc. This book is an entertaining excursion into mathematics, logic, and argument. Though it's dated in some ways, it's still worth digging out of the public library.

Moroney, M. J. 1953. *Facts From Figures*. Harmondsworth, Middlesex, UK: Penguin Books Ltd. Another book from long ago, it's one of the best I've seen at explaining the practice of statis-

tical analysis in the real world. It uses dozens of examples to illustrate statistical concepts, and while it can be technical at times, even lay readers will learn quickly from it.

11 • The Power of Critical Thinking

Hughes, William. 1997. *Critical Thinking: An Introduction to the Basic Skills*. 2d Peterborough, Ontario: Broadview Press. On pages 108-110 of this outstanding book, Hughes gives his seven rules for assessing arguments.

Hammond, John S., Ralph L. Keeney, and Howard Raiffa. 1999. *Smart Choices: A Practical Guide to Making Better Decisions*. Boston, MA: Harvard Business School Press. A methodical guide to thinking through your choices.

Mayfield, Marlys. 1994. *Thinking for Yourself: Developing Critical Thinking Skills Through Reading and Writing*. Belmont, CA: Wadsworth Publishing Company. A superb textbook that teaches clear writing and critical thinking using examples and exercises.

12 • Numbers Aren't Everything

Bella, David A. 1979. "Technological Constraints on Technological Optimism." *Technological Forecasting and Social Change*. vol. 14, p. 15. An article that uses the methods of science and mathematics to show why social and environmental problems generated by technology can be difficult to solve.

Jonas, Hans. 1973. "Technology and Responsibility: Reflections on the New Task of Ethics." *Social Research*. vol. 40, no. 1. p. 31. This article addresses ethical issues that affect modern technological societies that for the first time have the power to effect changes in the natural environment in a global way. As technology changes what it means to be human, Jonas argues that contemporary ethics must keep pace.

13 • All Numbers Are Not Created Equal

DeCanio, Stephen. 2003. *Economic Models of Climate Change: A Critique*. Basingstoke, UK: Palgrave-MacMillan.

Schwartz, Peter. 1996. *The Art of the Long View: Planning for the Future in an Uncertain World*. New York, NY: Doubleday. The classic book on scenario analysis. A "must read" book for anyone trying to think systematically about the future.

14 • Question Authority

Cialdini, Robert B. 1993. *Influence: The Psychology of Persuasion*. New York, NY: Quill–William Morrow. An extraordinary book that documents how "compliance professionals" influence people's opinions and actions.

Crossen, Cynthia. 1994. *Tainted Truth: The Manipulation of Fact in America*. New York, NY: Simon & Schuster. Another eye opener that shows how people and institutions can and do play with numbers to convince and confuse.

Hughes, William. 1997. *Critical Thinking: An Introduction to the Basic Skills*. 2d Peterborough, Ontario: Broadview Press. Hughes presents detailed criteria for assessing the strength of appeals to authority.

Robbins, John. 1987. *Diet for a New America*. Walpole, NH: Stillpoint Publishing. Whatever you think about Robbins' advocacy of a vegetarian diet, his description of how the dairy and meat industries got schools and other public institutions to distribute their marketing materials is a real eye opener.

16 • Don't Believe Everything You Read

Crossen, Cynthia. 1994. *Tainted Truth: The Manipulation of Fact in America*. New York, NY: Simon & Schuster. Crossen summarizes how the results of research are often manipulated for self-serving ends.

Ewen, Stuart. 1996. *PR! A Social History of Spin*. New York, NY: Basic Books, a member of the Perseus Books Group. This book recounts the development of public relations and propaganda in the Twentieth Century. It is an intriguing and well-told tale.

Huff, Darrell. 1993. *How to Lie with Statistics*. New York, NY: W. W. Norton & Co., Inc. If this book doesn't make you a skeptical observer of numbers and statistics, nothing will. First published in the 1950s, it's still fresh and useful today.

17 • Go Back to the Questions

Crossen, Cynthia. 1994. *Tainted Truth: The Manipulation of Fact in America*. New York, NY: Simon & Schuster. Crossen demonstrates how research has become a tool for hucksters to make friends and influence people.

Weiss, Michael J. 1994. *Latitudes and Attitudes: An Atlas of American Tastes, Trends, Politics, and Passions*. New York, NY: Little Brown and Company. This book is a beautiful summary of consumer and market data. It shows the power of comprehensive consumer surveys to describe consumer preferences across the U.S.

18 • Reading Tables and Graphs

Tufte, Edward R. 1995. *The Visual Display of Quantitative Information*. Cheshire, CT: Graphics Press. Fourteenth printing. Reading this book will change the way you look at tables and graphs, and it is required reading for those learning this craft.

19 • Distinguish Facts from Values

Booth, Wayne C., Gregory G. Colomb and Joseph M. Williams. 1995. *The Craft of Research*. Chicago, IL: The University of Chicago Press. This superb book presents a framework for dissecting arguments that can be applied in most situations. The authors were trained in rhetoric and literature, but their way of thinking about claims and evidence applies equally well to technical topics.

Fuchs, Victor R., Alan B. Krueger and James M. Poterba. 1997. *Why Do Economists Disagree About Policy? The Roles of Beliefs About Parameters and Values*. Princeton University, Industrial Relations Section. Working Paper #389. August.

Jonas, Hans. 1973. "Technology and Responsibility: Reflections on the New Task of Ethics." *Social Research*. vol. 40, no. 1. p. 31. This article, more than any other I read in graduate school, got me to think carefully about how technology can challenge some of our most deeply held values.

Marshall, Jonathan. 1997. "Economists Can't Seem to Agree on Anything." *The SF Chronicle*, August 25, 1997, p. B2.

Reich, Robert. 1997. *Locked in the Cabinet*. New York, NY: Alfred A. Knopf. Regardless of your political leanings, you'll enjoy Reich's tales of political intrigue, error, and innuendo in the Clinton White House.

20 • The Uncertainty Principle and the Mass Media

Cialdini, Robert B. 1993. *Influence: The Psychology of Persuasion*. New York, NY: Quill–William Morrow. Cialdini's book shows how "compliance professionals" use their knowledge of the human psyche to influence people's decisions.

Crossen, Cynthia. 1994. *Tainted Truth: The Manipulation of Fact in America*. New York, NY: Simon & Schuster. Crossen gives dozens of examples of how the results of research and reporting can be manipulated for good or ill.

Fuller, Jack. 1996. *News Values: Ideas for an Information Age*. Chicago, IL: University of Chicago Press. Fuller explores how new technologies continue to change the role of the journalist in modern society.

Levering, Robert, and Milton Moskowitz. 1994. *The 100 Best Companies to Work for in America*. Plume Publishing. This is the 1994 edition of their book, which was first published in 1984.

PART IV • CREATE YOUR ANALYSIS

21 • Reflect

Hyams, Joe. 1982. *Zen in the Martial Arts*. New York, NY: Bantam Books.

Mattimore, Bryan W. 1994. *99% Inspiration: Tips, Tales & Techniques for Liberating Your Business Creativity*. New York, NY: AMACOM (A Division of American Management Association).

Ueland, Brenda. 1987. *If You Want to Write: A Book About Art, Independence, and Spirit*. St. Paul, Minnesota: Graywolf Press.

22 • Get Unstuck

Elffers, Joost. 1997. *Play with Your Food*. New York, NY: Stewart, Tabori, Chang. This book encourages you to think about food in a way you never have before. If you really get stuck, get this book and play until you're ready to get back to the problem.

Mattimore, Bryan W. 1994. *99% Inspiration: Tips, Tales & Techniques for Liberating Your Business Creativity*. New York, NY: AMACOM (A Division of American Management Association). This book is a treasure trove of ideas for becoming more productive.

Polya, G. 1957. *How to Solve It: A New Aspect of Mathematical Method*. 2d ed., New York, NY: Doubleday & Company, Inc. A serious but approachable book about problem solving. It's a real classic!

23 • Inquire

Berkman, Robert I. 1990. *Find It Fast: How to Uncover Expert Information In Any Subject*. New York, NY: Harper & Rowe, Publishers. One researcher's tricks to digging out information in the real world.

U.S. Bureau of the Census. 1999. *The Statistical Abstract of the United States 1999*. 119th edition Washington D.C.: U.S. Government Printing Office. One of the most important data sources for any U.S. researcher.

Edwards, Paul, ed. 1972. *The Encyclopedia of Philosophy*. New York, NY: Macmillan Publishing

Co., Inc. & The Free Press. Volumes 1 through 8. This set is an essential desk reference that describes the historical development of ideas. It can lend context to your research available nowhere else.

The Encyclopedia Brittanica. The hard copy of this encyclopedia costs more than $1000, but the electronic versions are a much better deal. A multimedia CD-ROM version costs about $125 (for both Windows and Macintosh), and a subscription-based web version <http://www.eb.com> costs $8.50/month or $85/year. Other encyclopedias offer similar deals.

24 • Be a Detective

Sir Arthur Conan Doyle. *A Study in Scarlet.* This novel, in which Conan Doyle created the modern detective, first brought Sherlock Holmes into English literature. It's a gripping, well-told story, and one in which the scientific method features prominently. Contained in Baring-Gould, William S. ed. 1976. *The Annotated Sherlock Holmes.* Volume I. New York, NY: Clarkson N. Potter, Inc./Publisher. 15th printing. pp. 143-234.

25 • Create Consistent Comparisons

Koomey, Jonathan G., Deborah E. Schechter and Deborah Gordon. 1993. *Cost-Effectiveness of Fuel Economy Improvements in 1992 Honda Civic Hatchbacks.* Energy and Resources Group, University of California, Berkeley. Presented at the 72nd Annual meeting of the National Research Council's Transportation Research Board, January 10-14, 1993, Washington, DC. May 1993. Download the report in PDF format at <http://enduse.lbl.gov/Info/Honda-abstract.html>.

26 • Tell a Good Story

Morgan, M. Granger, and Max Henrion. 1992. *Uncertainty: A Guide to Dealing with Uncertainty in Quantitative Risk and Policy Analysis.* New York, NY: Cambridge University Press.

Schwartz, Peter. 1996. *The Art of the Long View: Planning for the Future in an Uncertain World.* New York, NY: Doubleday.

27 • Dig into the Numbers

Cook, Terry. 1995. "It's 10 O'Clock: Do You Know Where Your Data Are?" *Technology Review.* January. p. 48. An insightful article that discusses some unprecedented problems with data that we face in the information age.

28 • Make a Model

Keepin, William (Bill). 1983. *A Critical Appraisal of the IIASA Energy Scenarios.* International Institute for Applied Systems Analysis. WP-83-104. October. The complete documentation for Keepin's criticisms of the IIASA scenarios.

Keepin, William and B. Wynne. 1984. "Technical analysis of IIASA energy scenarios." *Nature.* vol. 312, p. 691. This is the journal article summary of Keepin's work at IIASA.

Morgan, M. Granger and Max Henrion. 1992. *Uncertainty: A Guide to Dealing with Uncertainty in Quantitative Risk and Policy Analysis.* New York, NY: Cambridge University Press. This book gives sage advice about appropriate uses for models in analysis, and is one of the few places where large complex models are taken to task for their failings.

Ringer, Robert J. 1974. *Winning Through Intimidation*. New York: Funk & Wagnalls. A funny and entertaining look at one man's transformation from loser to winner, in part through his clever use of models.

Starfield, Anthony M., Karl A. Smith and Andrew L. Bleloch. 1990. *How to Model It: Problem Solving for the Computer Age*. New York, NY: McGraw-Hill, Inc. This book is an excellent, albeit technical, guide to modeling practice in the real world, particularly computer modeling. It teaches problem solving skills essential for good modeling, and focuses mainly on scientific applications, though its approach is useful for non-scientists as well. It is unfortunately out of print, but many libraries will have it.

29 • Reuse Old Envelopes

Harte, John. 1985. *Consider a Spherical Cow: A Course in Environmental Problem Solving*. Los Altos, CA: William Kaufmann, Inc. The first short chapter in this classic book asks the question "How many cobblers are there in the U.S.?" and then proceeds to investigate ever more detailed puzzles in environmental science. This book, and the course in which it is used, were inspirational to me in graduate school. The sequel to it (*Consider a Cylindrical Cow: Further Adventures in Environmental Problem Solving*) was published in 2001 by University Science Books, in Sausalito, California.

Herendeen, Robert A. 1998. *Ecological Numeracy: Quantitative Analysis of Environmental Issues*. New York, NY: John Wiley & Sons, Inc. This book contains many nice examples of back-of-the envelope calculations, and gives a somewhat technical but still useful introduction to issues of energy and the environment.

Hershey, Robert L. 1982. *How to Think with Numbers*. Los Altos, CA: William Kaufmann, Inc. This book is hard to find, but if you're interested in learning about some relevant tricks, it's worth tracking down.

Hershey, Robert L. 1997. "Thinking Big." *The Washington Post*, September 10, 1997, p. H8. A short article talking about scientific notation and its usefulness for treating really BIG numbers.

Mattimore, Bryan W. 1994. *99% Inspiration: Tips, Tales & Techniques for Liberating Your Business Creativity*. New York, NY: AMACOM (A Division of American Management Association). "Chapter 10–Fermi Problems: Guesstimating in Today's Decimal-Point World."

Ross, Joan and Marc Ross. 1986. "Fermi problems, or how to make the most of what you already know." In *Estimation and Mental Computation: 1986 Yearbook*. Edited by H. L. Schoen and M. J. Zweng. National Council of Teachers of Mathematics.

von Baeyer, Hans Christian. 1993. *The Fermi Solution: Essays on Science*. New York, NY: Random House. von Baeyer's ruminations on science are interesting and illuminating. His chapter on Fermi problems discusses their history.

30 • Use Forecasts with Care

Armstrong, J. Scott. 2000. *The Forecasting Principles Web Site*, accessible at <http://www.forecastingprinciples.com>.

Bright, James R. 1972. *A Brief Introduction to Technology Forecasting: Concepts and Exercises*. Austin, Texas: The Pemaquid Press. This book is a bit dated in some ways, and its optimism about accurate forecasting is a bit quaint, but it's long on examples, practical advice, and careful thinking about forecasting in the real world.

Ehrlich, Paul R., Anne H. Ehrlich, and John P. Holdren. 1977. *Ecoscience: Population, Resources, Environment*. San Francisco, CA: W.H. Freeman and Company. A beautiful book with a nice

writeup of the subtleties of population forecasts, exponential growth, and logistic curves (see Chapter 4). Even though it's more than two decades old, there's still nothing like it.

Institute for Business Forecasting. <http://www.ibf.org>. email: ibf@ibf.org. 800/440-0499, 718/463-3914, 718/544-9086 fax. Flushing, NY. An institution devoted to disseminating knowledge about business forecasting and planning. If you're interested in forecasting in a business context, this organization is a good place to start.

Morgan, M. Granger, and Max Henrion. 1992. *Uncertainty: A Guide to Dealing with Uncertainty in Quantitative Risk and Policy Analysis.* New York, NY: Cambridge University Press. One of the best books on uncertainty ever written. A true classic.

Moroney, M. J. 1953. *Facts From Figures.* Harmondsworth, Middlesex, UK: Penguin Books Ltd. This statistics book has an excellent long chapter on correlation versus causality, and a wonderful short chapter on the perils of trend-based forecasting. My favorite quotation from Moroney: "Economic forecasting, like weather forecasting in England, is only valid for the next six hours or so. Beyond that it is sheer guesswork."

Schnaars, Steven. 1989. *Megamistakes: Forecasting and the Myth of Rapid Technological Change.* New York, NY: Free Press. This book gives examples of major forecasting failures. Both lively and insightful, this book is a great read for those interested in pitfalls in forecasting.

PART V • SHOW YOUR STUFF

General

Williams, Robin. 1994. *The Non-Designer's Design Book : Design and Typographic Principles for the Visual Novice.* Berkeley, CA: Peachpit Press. This book is a beautiful and useful introduction to design principles. Williams teaches the key concepts of proximity, alignment, repetition, and contrast, all of which are critical to presenting information effectively. She also reviews the use of typefaces.

Hacker, Diana. 1999. *A Writer's Reference.* Boston, MA: Bedford/St. Martins. A superb sourcebook dealing with the craft of writing, including grammar, spelling, punctuation, references and creating compelling arguments. The associated web site is also a great resource: <http://www.bedfordstmartins.com/hacker/writersref>.

32 • Know Your Audience

Fuller, Jack. 1996. *News Values: Ideas for an Information Age.* Chicago, IL: University of Chicago Press. A clearly written book that explores the changing relationships between journalists and their readers in the age of electronic media.

33 • Document, Document, Document

The Chicago Manual of Style: The Essential Guide for Writers, Editors, and Publishers. 1993. 14th ed. Chicago, IL: University of Chicago Press. The classic reference on style and formats, including discussion of spelling, punctuation, grammar, tables, abbreviations, numbers, foreign languages, and other complexities afflicting the modern writer.

Cook, Terry. 1995. "It's 10 O'Clock: Do You Know Where Your Data Are?." *Technology Review.* January. p. 48. An eye-opening article that discusses the real world problems with keeping track of data in the electronic age.

Hacker, Diana. 1999. *A Writer's Reference.* Boston, MA: Bedford/St. Martins. Part of the web site

associated with this book has information and examples on research and documentation in the electronic age, with separate guidance for sciences, social sciences, and humanities: <http://www.dianahacker.com/resdoc/>.

Stim, Richard. 2000. *Getting Permission: How to License & Clear Copyrighted Materials, Online and Off*. Berkeley, CA: Nolo.com. <http://www.nolo.com/>. Practical advice on getting permission to use copyrighted materials.

34 • Let the Tables and Graphs Do the Work

Booth, Wayne C., Gregory G. Colomb and Joseph M. Williams. 1995. *The Craft of Research*. Chicago, IL: The University of Chicago Press. One of the best books on the process of writing research papers, from start to finish.

Strunk Jr., William, and E. B. White. 1979. *The Elements of Style*, 3d ed. New York, NY: MacMillan Publishing Company, Inc. The basic *Elements of Style* hasn't changed in decades (it's one of those gems that simply can't be improved upon). Plan on rereading this short but essential book every few years, just to refresh your memory and enliven your writing.

The Chicago Manual of Style: The Essential Guide for Writers, Editors, and Publishers. 1993. 14th ed. Chicago, IL: University of Chicago Press. Many writers consider this book to be the last word in rules for writing. It's a comprehensive source, and one that's well worth having on your shelf.

35 • Create Compelling Graphs and Figures

Craver, John S. 1980. *Graph Paper From Your Copier*. Tucson, AZ: HPBooks. If you need a special kind of graph paper in a pinch, this book is just the ticket, with almost 200 different graph and map types, ready for copying and use. It also has a nice 20+ page writeup of hints for making good graphs. It was especially useful before computer graphing became so widespread, but it still can come in handy.

Jones, Gerald E. 1995. *How to Lie with Charts*. San Francisco, CA: Sybex. Mainly oriented towards users of computer graphing tools, this book contains dozens of specific recommendations for making effective use of charts and graphs.

Monmonier, Mark. 1996. *How to Lie with Maps*. Chicago, IL: University of Chicago Press. This book contains examples of maps that mislead, confuse, advocate, and advertise. If you work with maps a lot, it's a "must read."

Munter, Mary. 1987. *Guide to Managerial Communication*. 2d Englewood Cliffs, NJ: Prentice Hall, Inc. This book describes techniques for effectively using graphs in a management context.

Tufte, Edward R. 1990. *Envisioning Information*. Cheshire, CT: Graphics Press. Tufte says this book is about *"pictures of nouns,"* namely graphical representations of places, things, people, and animals. It also deals with "visual strategies for design: color, layering, and interaction effects."

Tufte, Edward R. 1995. *The Visual Display of Quantitative Information*. Cheshire, CT: Graphics Press. Fourteenth printing. Tufte says this book is about *"pictures of numbers*, how to depict data and enforce statistical accuracy." It is by far the best book from which to learn the subtleties of effective graph design.

Tufte, Edward R. 1997. *Visual Explanation: Images and Quantities, Evidence and Narrative*. Cheshire, CT: Graphics Press. Tufte says this book is about *"pictures of verbs*, the representation of mechanism and motion, of process and dynamics, of causes and effects, of explanation and narrative." It focuses on the use of graphics to explain events and make explanatory assertions about those events.

Zelazny, Gene. 1996. *Say It With Charts: The Executive's Guide to Visual Communication.* 3d edition. Chicago, IL: Irwin Professional Publishers. A fine practical guide to using graphs in the business world.

36 • Create Good Tables

Jones, Gerald E. 1995. *How to Lie with Charts.* San Francisco, CA: Sybex. Has a nice section on table design.

Munter, Mary. 1987. *Guide to Managerial Communication.* 2d Englewood Cliffs, NJ: Prentice Hall, Inc. A business-oriented guide to getting your point across.

Tufte, Edward R. 1995. *The Visual Display of Quantitative Information.* Cheshire, CT: Graphics Press. Fourteenth printing. Key design principles and examples of good tables are sprinkled liberally throughout this superb book.

Williams, Robin. 1994. *The Non-Designer's Design Book: Design and Typographic Principles for the Visual Novice.* Berkeley, CA: Peachpit Press. The design principles in this book (including its introduction to typefaces) are especially pertinent to designing good tables.

37 • Use Numbers Effectively in Oral Presentations

Jones, Gerald E. 1995. *How to Lie with Charts.* San Francisco, CA: Sybex. Gives specific advice on using tables and graphs in oral presentations.

Osgood, Charles. 1988. *Osgood on Speaking: How to Think on Your Feet Without Falling on Your Face.* New York, NY: William Morrow and Company, Inc. One of the best books on the art of public speaking.

Tufte, Edward R. 1997. *Visual Explanation: Images and Quantities, Evidence and Narrative.* Cheshire, CT: Graphics Press. In this book, Tufte derives lessons from magicians and illusionists to teach good presentation skills (see pp. 68-70).

38 • Use the Internet

Apple Computer. 1992. *Macintosh Human Interface Guidelines.* Reading, MA: Addison-Wesley Publishing Company. This book gives general information on how computers and humans can most peacefully coexist. The general principles it espouses are useful regardless of which operating system you favor. It is also available on-line at <http://developer.apple.com/documentation/UserExperience/UserExperience.html>.

Burke, James. 1978. *Connections.* Boston: Little, Brown and Company. An outstanding book. Among other things, Burke describes the key role of the printing press in fostering later technological and scientific developments.

Greenspun, Philip. 1999. *Phil and Alex's Guide to Web Publishing.* San Francisco, CA: Morgan Kauffman. While somewhat technical in parts, this is absolutely the best book for people interested in making the web useful. It's also a beautiful coffee table book. For the on-line version, see <http://philip.greenspun.com/panda/>.

Williams, Robin. 1994. *The Non-Designer's Design Book : Design and Typographic Principles for the Visual Novice.* Berkeley, CA: Peachpit Press. This book is a wonderful introduction to design principles, many of which apply directly to web design.

Williams, Robin and John Tollett. 1997. *The Non-Designer's Web Book : An Easy Guide to Creating, Designing, and Posting Your Own Web Site.* Berkeley, CA: Peachpit Press. A delightful and easy to read book that gives you the basics of web design and implementation, dealing with nitty-gritty details of file formats, typography, colors, effective design strategies, and all sorts of other stuff.

NOTES

1. Affiliations listed for informational purposes only. The opinions expressed do not necessarily represent those of the institutions listed.

2. I am indebted to Amory Lovins for this phrase.

3. Tufte, Edward R. 1995. *The Visual Display of Quantitative Information*. Cheshire, CT: Graphics Press. Fourteenth printing. p. 51.

4. By the early 1990s, the LoC collection totaled more than 100 million items, of which about half are books and manuscripts <http://lcweb.loc.gov/loc/legacy/colls.html>. For this calculation I assume that eighty percent of the books and manuscripts are duplicate copies of existing documents, so that our hapless reader would only have to peruse 10 million volumes.

5. An "order of magnitude" is a factor of ten, two orders of magnitude is a factor of one hundred, and so on. This shorthand is commonly used by scientists to summarize large differences in a convenient way.

6. More than half of all scientists who have ever lived are alive today (Stephen Covey, 1992. *Principle-Centered Leadership*. NY, NY: Simon & Schuster. p.91.), and this fact is reflected in ever more rapid rates of scientific and technological progress.

7. Shenk, David. 1997. *Data Smog: Surviving the Information Glut*. New York, NY: Harper-Collins Publishers, Inc. The first two paragraphs of this quotation appeared on p. 30 of *Data Smog*. The third paragraph appeared in various electronic versions of Shenk's work, but did not appear in the book itself.

8. In one survey, 80 percent of respondents "felt driven to gather as much information as possible, but over half of them were unable to handle all the information they accumulated." "Survey: New 'dataholics' generation on rise." *San Jose Mercury News*. 8 December 1997. Read on the web at <http://www.sjmercury.com>.

9. Ringer, Robert J. 1974. *Winning Through Intimidation*. New York: Funk & Wagnalls.

10. Hess, Peter, and Julie Siciliano. 1996. *Management: Responsibility for Performance*. New York, NY: McGraw-Hill, Inc. See Chapter 4: Developing goals.

11. Kawasaki, Guy. 1995. *How to Drive Your Competition Crazy: Creating Disruption for Fun and Profit*. New York, NY: Hyperion. p. 46.

12. Kawasaki, Guy. 1995. *How to Drive Your Competition Crazy: Creating Disruption for Fun and Profit*. pp. 40-41.

13. Kawasaki, Guy. 1995. *How to Drive Your Competition Crazy: Creating Disruption for Fun and Profit*. pp. 192-193.

14. Coffee, Peter. 1997. "Don't get sidetracked by trifling differences." *PC Week*. 24 November 1997. p. 32.

15. Hansell, Saul. "In terms of the audience, size matters." NYT May 11, 1998. Read on the web at <http://www.nyt.com>. The Audit Central site, launched in 2000 (but now defunct), standardized the ranking process for sites, so that they could be compared consistently.

16. Reich, Robert. 1997. *Locked in the Cabinet*. New York, NY: Alfred A. Knopf. p. 73.

17. Holton, Gerald. 1973. *Thematic Origins of Scientific Thought: Kepler to Einstein*. Cambridge, MA: Harvard University Press.

18. One famous example is that of Lord Kelvin's estimate of the age of the earth in the mid to late 1800s. Joe D. Burchfield, 1990. *Lord Kelvin and the Age of the Earth*. Chicago, IL: University of Chicago Press. For accounts of dozens of such mistakes, see Youngson, Robert. 1998. *Scientific Blunders: A Brief History of How Wrong Scientists Can Sometimes Be*. New York, NY: Carroll & Graf Publishers, Inc.

19. Significant effort has been made (and this effort extensively documented) to find the reproduction rights holder for Figure 4.1. If that person or institution contacts us with documentation of their ownership, we will negotiate suitable remuneration for use of the figure.

20. The general principles can be derived using either deductive or inductive logic.

21. Edwards, Paul, ed. 1972. *The Encyclopedia of Philosophy*. New York, NY: Macmillan Publishing Co., Inc. & The Free Press. Volume 4, p. 169.

22. Edwards, Paul, ed. 1972. *The Encyclopedia of Philosophy*. Volume 4, p. 169.

23. Sokal, Alan. 1997. "A Plea for Reason, Evidence and Logic." *New Politics* vol. 6, no. 2, pp. 126-129 (Winter 1997).

24. I am indebted to Bill Ahern for prompting me to ask these questions of myself.

25. Covey, Stephen R. 1990. *The 7 Habits of Highly Effective People*. NY, NY: Simon & Schuster. pp. 150-156.

26. Schwartz, Peter. 1996. *The Art of the Long View: Planning for the Future in an Uncertain World*. New York, NY: Doubleday. p.62.

27. In fairness, I made my share of wild assertions in those days.

28. Covey, Stephen R. 1990. *The 7 Habits of Highly Effective People*. NY, NY: Simon & Schuster. pp. 150-156.

29. DeMarco, Tom, and Timothy Lister. 1987. *Peopleware: Productive Projects and Teams*. New York, NY: Dorset House Publishing Company. p. 55.

30. Hughes, William. 1997. *Critical Thinking: An Introduction to the Basic Skills*. 2d edition. Peterborough, Ontario: Broadview Press. p. 16.

31. Hughes, William. 1997. *Critical Thinking: An Introduction to the Basic Skills*. pp. 108-110.

32. I am indebted to Professor John Holdren for this phrase.

33. Quoted in Stobough, Robert, Daniel Yergin, I. C. Bupp, Mel Horwitch, Sergio Koreisha, Modesto A. Maidique and Frank Schuller. 1979. *Energy Future*. New York, NY: Ballantine Books.

34. Hughes, William. 1997. *Critical Thinking: An Introduction to the Basic Skills*. pp. 160-161.

35. Meier, Alan. 1987. "Saving the 'Other' Energy in Homes." *Energy Auditor and Retrofitter*. November/December, p. 13.

36. US DOE. 1989. *Residential Energy Consumption Survey (RECS): Housing Characteristics 1987*. EIA, Energy Information Administration, U.S. Department of Energy. DOE/EIA-0314(87). May. <http://www.eia.doe.gov>

37. No matter how rough they may be, Alan's estimates are based on years of experience and professional judgment. They prompt further empirical work that more often than not supports his initial guesses.

38. Calwell, Chris J. 1996. *Halogen Torchieres: Cold Facts and Hot Ceilings*. Boulder, CO: E-Source, Inc. TU-96-10. September.

39. If you subscribe to the *New Republic*, you can read the letter of apology at <http://www.tnr.com/archive/0698/060198/ourreaders060198.html>.

40. For another nice discussion of some common statistical fallacies, check out Barnett, Arnold. 1994. "How numbers are tricking you." *Technology Review*. October. <http://www.technologyreview.com/>

41. 2 Campus Boulevard, Newtown Square, PA 19073, 800/442-0925.

42. I am told that ants use a scented trail to show the way back to the nest. Any scented product (like talcum powder) will confuse them and make them go away.

43. Udell, Jon. 1996. "Web Surveys: Helpful Techniques for Web-based Data Collection and Analysis." *Byte Magazine*. October, p. 133.

44. Crossen, Cynthia. 1994. *Tainted Truth: The Manipulation of Fact in America*. New York, NY: Simon & Schuster, p. 112.

45. Cohen, Bernard L. 1990. *The Nuclear Energy Option: An Alternative for the 90s*. New York, NY: Plenum Press.

46. Reich, Robert. 1997. *Locked in the Cabinet*. New York, NY: Alfred A. Knopf, p.106.

47. Marshall, Jonathan. 1997. "Economists Can't Seem to Agree on Anything." *The SF Chronicle*, August 25, 1997, p. B2.

48. This figure is adapted from "Strategies, Structures, and Processes of Organizational Decisions," which is from *Comparative Studies in Administration*, T. E. Thompson and A. Tuden, (c) 1959. University of Pittsburgh Press.

49. Kunde, Diana. 1998. "Don't Be Bugged." *The SF Chronicle*. Sunday, March 8, 1998, p. J1.

50. Crossen 1994, *Tainted Truth*, p. 106.

51. Crossen 1994, *Tainted Truth*, p. 94.

52. Crossen 1994, *Tainted Truth*, p. 101.

53. Dougan, Michael. 1996. "Ancient quest tests new library." *Sunday SF Examiner*. December 15, 1996, p. C1.

54. For a nice introduction to the concept of a spreadsheet, see Jones, Gerald E. 1995. *How to Lie with Charts*. San Francisco, CA: Sybex, in the section titled "Spreadsheets: The Crash Course," on pages 154-157.

55. CNET online retells the story of this launch as well as other famous computer glitches in "CNET Features—Digital Life—Bugs" (8/24/2000).

56. Sawyer, Kathy. 1999. "Engineers' Lapse Led to Loss of Mars Spacecraft: Lockheed Didn't Tally Metric Units." *The Washington Post*. October 1, 1999, p. A1.

57. Aragon, Lawrence. 1997. "Down with Dirt." *PC Week*. October 27, 1997, p. 83.

58. US raw steel production from 1997 *World Almanac and Book of Facts*, p. 153.

59. U.S. population from *1997 World Almanac and Book of Facts*, p. 382.

60. 1940-1980 GNP from the *Statistical Abstract of the US 1990*, p. 425. 1990 GNP in current $ from *1997 World Almanac and Book of Facts*, p. 133, adjusted to 1982 dollars using the consumer price index from p. 132.

61. The easiest way to calculate this number is to take the ratio of market capitalization per dollar of annual revenues for GE, and divide by the same ratio for Microsoft. This calculation yields 0.12, which equals 1/8.

62. From the perspective of actually *using* the data for calculations (as in this example), downloadable files are almost always preferable to such HTML pages (see Chapter 38: Use the Internet).

63. Ringer, Robert J. 1974. *Winning Through Intimidation*. New York: Funk & Wagnalls.

64. Morgan, M. Granger, and Max Henrion. 1992. *Uncertainty: A Guide to Dealing with Uncertainty in Quantitative Risk and Policy Analysis*. New York, NY: Cambridge University Press.

65. "Trickle-Down Plan." *The SF Chronicle*. December 5 1996, p. E10.

66. This calculation highlights a startling fact: gasoline is far cheaper than bottled water or almost any other purchased fluid you can name, in spite of its extraordinary energy density. For a nice graph demonstrating this result, see Nadis, Steve and James J. MacKenzie. 1993. *Car Trouble*. World Resources Institute. Washington, DC.

67. Currier, Chet. *The SF Chronicle*. November 25, 1996, p. E3.

68. *The San Francisco Chronicle/Examiner*. Sunday, August 31, 1997. Housing prices were examined on p. E1, and cremation was examined on p. D1.

69. I am indebted to Mark Nowak for this marvelous anecdote.

70. Such intellectual debate is often inhibited between companies because of concerns about proprietary information. These concerns can sometimes be allayed by appropriate use of non-disclosure agreements. Surprisingly, lack of constructive intellectual criticism is the norm even *within* firms, where proprietary concerns are moot.

71. Thanks to Veronica Oliva for her thoughts and comments on this section.

72. "Stamp Price to Rise a Penny to 33 Cents." *The SF Chronicle*. May 12, 1998, p. A1+.

73. Knutson, Lawrence L. 1998. "U.S. to Rain Pennies on Post Office: 33-cent First-class Stamp Takes Effect Next Year." *The Oakland Tribune*. May 12, 1998, p. NEWS-1+.

74. I am indebted to Professor Marcela Bilek for this point. She discovered the importance of it when examining historical trends in Japanese GDP, expressed in dollars. You can investigate for yourself at <http://www.economagic.com>.

75. Herhold, Scott. 1998. "Computer Files Erased at Stanford." *San Jose Mercury News*. April 9, 1998, p. 1A.

76. Tufte, Edward R. 1995. *The Visual Display of Quantitative Information*, p. 51.

77. Tufte, Edward R. 1995. *The Visual Display of Quantitative Information*, p. 170.

78. Vorsatz, Diana, Leslie Shown, Jonathan G. Koomey, Mithra Moezzi, Andrea Denver, and Barbara Atkinson. 1997. *Lighting Market Sourcebook*. Ernest Orlando Lawrence Berkeley National Laboratory. LBNL-39102. December. p. 38. <http://enduse.lbl.gov/Projects/LMS.html>.

79. Tufte, Edward R. 1995. *The Visual Display of Quantitative Information*, p. 191.

80. Tufte, Edward R. 1995. *The Visual Display of Quantitative Information*, p. 191.

81. Trefil, James. 1994. *A Scientist in the City*. New York, NY: Doubleday, p. 24.

82. Vorsatz, Diana, et al. 1997. *Lighting Market Sourcebook*. p. 51.

83. Interlaboratory Working Group. 1997. *Scenarios of U.S. Carbon Reductions: Potential Impacts of Energy-Efficient and Low-Carbon Technologies by 2010 and Beyond*. Oak Ridge National Laboratory and Lawrence Berkeley National Laboratory. ORNL/CON-444 and LBNL-40533. September. p. 2.12. <http://enduse.lbl.gov/Projects/5Lab.html>

84. I am particularly indebted to Evan Mills for his detailed comments and suggestions on this chapter.

INDEX